The Science of God Volume 1

The Science of God
The First Four Days

R Lindemann

Aleph Publications
Wisconsin, USA

The Science of God Volume 1
The First Four Days
Copyright 2015 - R Lindemann ©
All Rights Reserved. Published 2023

Aleph Publications
Manitowoc, WI

Paperback Edition
ISBN13: 978-1-956814-24-8

33 32 31 30 29 28 27 26 25 24   2 3 4 5 6

Disclaimer

All information, views, thoughts, and opinions expressed herein are those of the author(s) and are being presented only for your consideration and should not be interpreted as advice to take any action. Any action you take with regard to implementing or not implementing the information, views, thoughts, and opinions contained within this published work is your own responsibility. Under no circumstances are distributor(s) and/or publisher(s) and/or author(s) of this work liable for any of your actions.

Anyone, especially those who have been victim of misdirected explanation and understanding, may be best served seeking wise counsel before deciding to implement any information, views, thoughts, opinions, or anything else that is offered for your consideration in this work. All information, views, thoughts, and opinions in this work are not advice, directive, recommendation, counsel, or any other indication for anyone to take any action. All information, views, thoughts, and opinions offered herein are offered only as suggestions for your personal consideration, which is done of your own free will. Your life is your own responsibility; use it wisely.

Any use of trade names or mention of commercial sources is for informational purposes only and does not imply endorsement or affiliation.

Please note that most of the items in quotes in this book are from various versions of the Bible and may have been paraphrased.

Dedication

This book is dedicated to all who seek the truth whether you are a religious person bound to your God or a science person bound to your science. To all those who desire to believe in a God but have been outwitted by science's explanations of our existence, or if you don't believe in God and are frustrated by the apparent ignorance of the Christians that just don't seem to get it, and if you are willing to set aside your preconceived thoughts and always desire to seek and find the actual truth, then this book is dedicated to you!

Contents

Acknowledgements

I would like to take this opportunity to offer many thanks to all who have discussed these topics with me and to all those who have spent years pondering such questions as are discussed in this book. Even though many have now passed on, they have left their thoughts written for us all to ponder and learn from, and also to use as a type of roadmap of possibly where to look and of what to avoid.

And special thanks to those who offered their skills in the editing and discussions process of this book. Also, thank you to the many who have discussed such topics, but do not even realize how they might have contributed through the many discussions that have occurred over several decades preceding the compiling of this book and its counterparts in *The Science Of God* series.

And I would also like to acknowledge everyone, from the people who make the equipment that makes the tires on the equipment that cuts the trees that makes the paper that hard copies of this book are printed on, along with all of the people on every step of the way to the delivery person that hands you the package, and so many more people who are connected to all of that. Without this vast network of wonderful, hard-working people who share their talents with the world little good in this world would ever be accomplished. Thank you all!

Introduction

The debate about Creation and how everything came to be has been going on for a very long time. Many people are under the impression that it is only in our modern era that this debate has been a central discussion in astrophysics or even in science in general. Going way back to the so called "Age of Reason", these questions came to the forefront of discussion, but the debate is far older than that. These questions go back at least as far as the Greek philosophers and probably much further. People have likely been questioning how it all came to be since we humans came to be.

People often assume that the only debate is a spontaneous-big-bang versus the six-day-creation-by-god argument, and while in many cases this may be true now, it was not necessarily the situation thousands of years ago. The primary debate has nothing to do with if there is a God or if that God Created all things. The initial Creation debate is "how did it occur?" regardless of a Creator. That is to say that whether or not there is a God, in what manner did Creation occur? Was there a big bang? Or was everything always there? Or did the elements just sort of materialize? Or... what? Humans have been pondering this question generation after generation after generation, and truth be told, we will likely continue to do so until we die.

But setting aside the issue of questioning what method may have occurred for all things to exist, we also have that lingering question of, did it occur of its own accord, or did something or a "God" deliberately cause it to occur?

Separating the *how* and the *why* aspects of the questions regarding Creation offers you a clearer view of the larger picture, but if you are not able to separate the *how* and the *why*, then you have little hope of ever being able to understand any of it properly. One aspect is *how* it occurred and the other aspect is *why* it occurred. *How* versus *Why*. If there is no God then the *why* is irrelevant and only the *how* matters. However, when we toss out the idea of a God, then we end up with one primary unexplainable question, and that question is: What caused it to begin with?

Unfortunately, with regard to God, far too many people ignore sound-reasoning and attribute everything to God in a hocus-pocus like manner. In other words, God said "let there be..." and, POOF! Lights in the sky! And again, POOF! A ball of dirt called Earth! and so on. But is that sound reasoning? Is that what millennia of looking into the heavens and pondering it all, along with all of our years of collective scientific data reveal? Is that what actual real-world physics shows us?

That is the quest in *The Science of God* series exploring the major Biblical events; events that are often scientifically debated, but very poorly so. Events such as fundamental Creation or the flood are hotly contested topics and are often misunderstood by secular science, as well as by Biblical scholars. Over the years, I have found that there is too much focus on too few minds. We know this when we hear abused terms such as "follow the science" when there is no scientific method being implemented. As you read though this book, try to set aside your prejudicial knowledge and think openly. I urge you to consider the concept of "thinking openly" not as decrying religion and six-day creation, or for that matter denouncing science, but rather, approach these subjects with an inquisitive heart in an attempt to seek out the truth of what actually occurred regardless of what you were taught or want to believe.

Our preconceptions about such topics have been borne in us over our entire lifetime. It's difficult for us to see through the fog of what we unconsciously have chosen to believe due to the only exposures we have experienced. For instance, if you have never heard about Theory-C then you might believe that Theory-A and Theory-B are the only options, and so you will not even attempt to consider that a Theory-C exists at all.

Too often people feel alienated by religion for their own reasons, thus causing them to reject any idea of a God whatsoever, where on the other hand, others might see anomalies in science and outright reject science itself. Then there are those who will cling to God and refuse to see anything that does not agree with their own interpretation of events leading to and including Creation. While yet others cling to science in that same manner.

Always be open to finding the truth about Creation. When the actual truths are finally revealed, society will judge your beliefs accordingly, and who knows? Maybe your choices of belief will be tested at the pearly gates, if those gates actually exist. Let hardened positions not be yours. Be open enough to let Truth into your heart.

For years young Christians have attended college standing firm in their faith only to be crushed by their professors because the student did not understand and was only repeating what they thought they had been told. This then causes them to turn from the faith of their youth and eventually turn against God for the sake of their own vanity and self-protection. Then there are those who think they have it all figured out because to them the big bang explains everything, yet when you ask them deeper questions, they are unable to offer any useful answers. Both groups are "believers" of their respective religion. These same people will be the adults of tomorrow who will spread erroneous and vague information to yet another unsuspecting generation of unknowing youth.

Let us all strive for *accuracy* and *understanding* so that we may advance the understanding of all mankind.

Chapter 1

In the Beginning

"In the beginning God Created the Heavens and the Earth." Whoa, whoa, whoa! Wait a minute! Let's stop right there and analyze that statement, it is the very first sentence in the Bible and already we have a problem. Did you know that there are several versions of the Bible using somewhat different language *in the very first sentence* of the Bible?

What if there is no God? Then what created the "heavens" and the "earth"? Was it a big bang? Was everything packed into a tiny little ball that was so small that it couldn't even be scientifically detected, as big bang "singularity" proposes?

Is the Bible just a bunch of stories, or could it be an accurate account of Creation? And what are the "Heavens" and what is the "Earth"? We'll get into some of the more scientific aspects of Creation later, but first let's examine the text that is central to the debate between science and the Bible's "six day" account of Creation.

Which Bible version should we use to study with when we are asking specific questions? Which is the most accurate version? To answer those questions, we have to consider that every word spoken from one person to another is being translated or interpreted. The speaker is saying one thing and the listener must interpret the words of the speaker, and in the example just mentioned about every spoken word being translated, the speaker and the listener both fluently speak the same language. If you question this, then simply recall any time you have ever misunderstood someone's instructions or they have misunderstood your instructions. Now add to that the fact that the Bible was translated from one language to another. Additionally, with some Bibles there is more than one language-to-language interpretation in order to arrive at the modern versions that we read today.

In addition to all of that, we are reading the translator's concept of what was being conveyed in the text that the newly completed text was translated from to begin with. Then further, we must interpret those new words through the filter of our own preconceived ideas of *what we thought* we have been taught by other people.

Now, I assume that some readers will say "Ah Ha! I knew it! We just can't trust the Bible because there are too many translations, so how do we know which is the most authentic or closest to the original writing?" It all sounds a bit messed up, but don't get caught up in the nonsensical belief that the Bible is all messed up and altogether inaccurate. There is a great deal more on the translations and versions problem explained in the book *Understanding The Bible - The Bible How-To Manual* AND *The Things We Don't See.*

Don't let the differences in the Bible versions sway you into thinking that somehow they are all just a bunch of made-up stories or are somehow not accurate. When you review the most antiquated texts, you will find that there are obvious interpretation *similarities* and *differences*. In other words, the people who translated did their best in trying to convey their own understanding of each phrase that they translated. There is no other choice when translating any text. But the interesting part is that based upon the bulk of the Bible's words, most of the older versions dating from the 1500s and 1600s are very similar in almost all aspects. It is only in the more recent centuries that things get off track. And in this book, we need only concern ourselves with Genesis One.

We all must make sure that any information, views, thoughts, and opinions we are considering are only for **our own** consideration. It doesn't matter where you get your information, whether in this book or anywhere else, *you* are still ultimately responsible for what you choose to believe and follow, and that includes the Bible(s) you choose to use for study. That's why it's best if you consider the various thoughts and weigh the presented facts and information to draw *your own* conclusions.

Most people who dive into the deep-end of the Creation topic understand that there are fundamental Bible versions and texts which *most modern versions* are ultimately derived from—they are the Hebrew Texts, Greek Septuagint, the Latin Vulgate, and the Masoretic Text. When most Bible versions are compared against the more antiquated editions of these key versions, we find that they are mostly in unison even though there are possibly a few generations of versions between those versions and what we arrive at with some of today's current versions that most people are more familiar with.

As previously mentioned, the most antiquated of the key Bible versions are very much in unison. Yet, even that is argued by many so-called "Bible scholars". Debate often arises as to which version is best, and those opinions are dependent upon birth roots or chosen religious roots of each opinion holder. There is no question that the Bible versions have been, are now, and will be in the future, hotly debated by so-called "experts". But do not let their ignorance stop you from seeking the truth on your own.

When the specific text matches perfectly in the key versions then there should be little or no debate as to its authenticity.

The Concept of Translation

To obtain the best understanding of the Bible's Genesis One text, we first need to understand a little bit about the various Bible translations that we might come across while studying and scrutinizing the Genesis One text. There are good translations, but there are also bad translations that can cause problems when asking the tougher questions. Without getting too into the many details of the various Bible translations, the basics are that some of what we refer to as "The Bible" was translated into Greek roughly around 250 BC, into what we refer to as the "Septuagint". As far as is known, the Septuagint source documents were likely Hebrew and a few other texts available to the translators at that time. Some of the translated documents have been removed in recent centuries from most Protestant Christian Bible versions because they consider some writings to be "non-canonical".

In 382 AD the Bible was commissioned by Pope Damasus I to be translated into Latin by Jerome. Jerome's translation is commonly referred to as the "Vulgate". The Vulgate was translated using Hebrew, Aramaic, Old Latin, and Greek sources. There was a push for defining a standard assembly of these writings with regard to which books should be included in the Bible. This push was due to poor translations and false and/or very questionable gospels being injected into the early Christian movement. These questionable gospels had little proof of legitimacy and contradicted some of the key principles set down by God and by The Christ. Another critical reason that some books were excluded is because they were badly fragmented with important sections of text missing. Books with missing text can be dangerous when in the hands of the wrong people because those people can choose to fill in the missing information with fiction, thus deceiving those who are foolish enough to listen to their invented inaccuracies.

Then during its deliberations around 1550 AD, the Council of Trent made the Vulgate the official example of the Biblical canon and commissioned a standard version taken from Jerome's version now known as the Latin Clementine Vulgate, which was completed in the 1590s. The Clementine version was created to better format the text and to organize any added commentary found in the margins for a more concise display, ultimately making it easier to read.

What many people today do not realize is that if you were a Christian prior to the Reformation revolt era, then you were very likely considered a Catholic (Catholic means "universal", or "of the whole". See the book *Understanding The Church - Upon This Rock I Will Build My Church*). As is common with us humans, we often bend our better judgement for our own personal benefit, and some Church leaders were, or still are, no exception. Around the same period of time that Martin Luther was studying to be a Catholic priest, unrest was developing amongst the Church's vast congregation as the people pressed the issue to reform the Church.

The unrest regarding the actions and direction of some Church leaders was the catalyst for major reformation actions taken by some clergy, congregation, and political leaders. Martin Luther protested by posting his ninety-five theses on the church door in 1517 AD eventually leading to revolt. This revolt ultimately brought about the Lutheran religion as well as the Church of England or the Anglican Church in the 1530s AD due to the Catholic Church's refusal to grant an annulment to King Henry VIII. For good or for bad, this division has been an area of contention ever since the 1500s.

Consistency problems with the various books of the Bible existed to some extent prior to Jerome's 382 AD version, which is why the Canon was defined by the council and then later translated and compiled by Jerome into the single work similar to what we see today.

After the Reformation period began, the schisms between the reformers and the Catholic church clergy continued. Then partly due to spite and partly on a quest for accuracy, new translations of the Bible were created. One translation was done by Martin Luther into the German language around 1522 AD. Luther used Hebrew and ancient Greek and was also influenced by the Jarome Vulgate as source material for his translations. Another translation commissioned by King James was completed in 1611 AD. The English King James Bible translation used Greek, Hebrew, Aramaic, and Latin source material. In 1582 the Rheims New Testament English translation from Jerome's Latin Vulgate was completed, and then from the same source, in 1609 the Douay Old Testament English translation was completed. The earliest Masoretic text is believed to have been compiled from Jewish tradition, and from text possibly carried for centuries, though the information on this is somewhat ambiguous. The earliest texts are not specifically known. The oldest available Masoretic manuscript dates from around 900 AD.

There is much more detail to the history of the translations, but the information given here, while not extensive, should give you a roundabout idea of the basic flow of events leading to today's translations. If you want to know more about the Bible's assembly and the translation topics you can find much more information in the book *Understanding The Bible - The Bible How-To Manual* AND *The Things We Don't See* where this subject is discussed in further detail.

Here is an important point to consider about the various translations: When translated into English, the older versions mentioned here have nearly identical wording in Genesis One's account of Creation. And this is true of most translations that were done up until the late 1700s.

Any thought that the Bible has been greatly altered through many translations is simply false. In reality, there have been very few language-to-language translations to bring us the Bibles we read today. When those early translations were done they were meticulously scrutinized for accuracy, but some of the recent post-Reformation Bibles are not well translated.

The primary point of defining the Canon was to stop faulty versions from infiltrating into the Church. It has been greatly beneficial, while still contentious, for these various translations to be done separately. This fork of translation caused an inadvertent check system to be created, thus showing a tremendous amount of stability in the text and the understanding of that text, which is proven when you compare the text between the modern English translations of the Douay Rheims, King James, Luther Bible, and Masoretic Text. Make no mistake about it, there are other translations, but the same holds true, they all are in near perfect harmony regarding the Genesis One Creation account when they have not taken liberties with the text.

Ancient languages that are written down cannot change over the course of time because they are recorded in the very writing that defines them. Even today, languages such as old Latin are used in defining various scientific terms because the writing cannot change like the meanings of words tend to do in modern languages. Consider for a moment the atrocities that occurred in the modern English language around the beginning of the twenty-first century. It makes one shudder and fear for the upcoming generations if the perversion of the language fails to cease.

In general, the various books of the Bible have been accurately carried through thousands of years in various forms regardless of which antiquated version you read. For most people, the primary concern of the Bible is that of Salvation, which is that you cannot be saved without accepting The Christ's Death and Resurrection. The Masoretic Text is a Jewish version and the Jews have chosen to not accept the Christian Jesus as the promised savior, so the Masoretic text does not include the New Testament texts. In this book, *The Science of God Volume 1*, the Resurrection is not at all a part of the topic, so in this book we do not need to concern ourselves with the fact that the Masoretic version does not include the New Testament.

We need only concern ourselves with these key versions. But again, we are caught in a trap because those key versions of antiquity have been translated from their native tongue into modern English. You would think that if two people translated the identical text that the translations would be identical, but that is not the case in many translations printed since the eighteenth century.

Don't get me wrong here, most of the words translated are done to a point where most versions' Genesis One agree to near perfection. However, there are critical translation errors we find on a few key words in Genesis One in a few of these more recent translations. Do not let this concern you or confuse you. There are plenty of sources to get facsimiles of the 1500s and 1600s copies and you will find those to be very well translated compared to our modern English versions of them. If you plan to study and are getting into the deeper questions and your study Bible uses words such as "expanse", "canopy", "vault", "sky", or "horizon" rather than the word "firmament" in Genesis One, then set it aside and find a more authoritative Bible.

Since this book only involves the first four days of Creation, skipping most of day three, we only need to consider the first several verses of *any* Bible translation in the discussion regarding the astrophysical aspects discussed in this volume of *The Science Of God*. We are going to get a bit more particular about the grammar used during the description of Creation in the Bible in Genesis One. We will also discuss the importance of the differences and how those differences might negatively affect each our own understanding of the Genesis One text.

Scientists have an easy route, in that if they happen to be atheists they have no need of considering the Bible whatsoever in their conclusions and can invent any theory they can dream up.

It is sort of a coward's way out of making sense of all of the data we scientifically collect.

However, if you are a so-called "believer" and the Bible is your roadmap to life, then you have a much more difficult task ahead of you. If you intend on being viewed as "credible" by your fellow creationist or even by scientists, then you have to match scientific findings with what you believe the Bible says about Creation. And then do so without making God out to be some sort of hocus-pocus magician in the sky. And that is a task of monumental proportions when you do not properly grasp the Genesis One text.

A *believer* that has an open mind must find the truth, and the Bible must agree with science and vice-versa, or you will appear as a fool to many people. When considering these points and when reading through the various versions of the Bible, we can see some subtle differences that cause problems in the very first sentence of the Bible's Genesis Chapter One Verse One.

Take a look at some of the various English translation versions of the Bible's first sentence. There are many Bible versions and most are ultimately derived from the oldest Hebrew text, Septuagint, Masoretic, and Vulgate versions already mentioned. Here are some common Bible versions you might find in many homes, and following is their interpretation of the first sentence of Genesis One.

Douay-Rheims Bible (Largely translated from Septuagint and Masoretic)
    1)   In the beginning God created heaven, and earth.

JPS Tanakh 1917 – (Largely translated from Masoretic)
Webster's Bible Translation
English Revised Version
King James Bible – (Largely translated from Masoretic)
American King James Version – (Largely translated from King James Bible)
King James 2000 Bible – (Largely translated from King James Bible)
    2)   In the beginning God created the heaven and the earth.

New American Standard Bible
Holman Christian Standard Bible

NET Bible
New Living Translation
Jubilee Bible 2000
American Standard Version
Darby Bible Translation
World English Bible
New American Standard 1977
    3)   In the beginning God created the heavens and the earth.

GOD'S WORD Translation
    4)   In the beginning God created heaven and earth.

Young's Literal Translation
    5)   In the beginning of God's preparing the heavens and the earth.

English Standard Version
    6)   In the beginning, God created the heavens and the earth.

International Standard Version
    7)   In the beginning, God created the universe.

Now compare them without the text's Bible source-name interrupting the comparisons:

1) In the beginning God created heaven, and earth.
2) In the beginning God created <u>the</u> heaven and <u>the</u> earth.
3) In the beginning God created <u>the</u> heavens and <u>the</u> earth.
4) In the beginning God created heaven and earth.
5) In the beginning of God's <u>preparing</u> <u>the</u> heavens and <u>the</u> earth.
6) In the beginning, God created <u>the</u> heavens and <u>the</u> earth.
7) In the beginning, God created <u>the</u> <u>universe</u>.

Take note of the use of commas and plurals and the use of time inferred in the term "preparing", and the scope, of Creation with the term "universe", and also the use of a specific article via the use of the word "the".

Could any of these various versions influence someone's interpretation of the text while they are pondering how everything came to be? When you initially think about it, this might seem as if it is irrelevant, but when you start asking more detailed questions about the scientific aspects of Creation, then the nuances of the first sentence of the Bible and the sentences that follow carry much more weight in the analysis of the Creation account. And therefore, each word used can greatly influence a person's interpretation of subsequent sentences or events stated in the Creation account.

One of the biggest problems we face is each our own willingness to question things *properly*. On the atheist side, I often hear people utterly disregarding the Bible as if it is all a randomly made up story. On the Believer side, I hear people repeating the false preachers that they follow who have little or no knowledge of astrophysics, yet they espouse views of Creation that are childish. Neither one of these two approaches are useful to any one of us, nor do they bring us any closer to truth.

Included in the next several pages are facsimile versions of Genesis One dating from the 1500s and 1600s in the native language of the translation, with little effort you can find such facsimiles in digital format in our modern era:

Jerome's Vulgate

IN principio creavit ds
  caelum et terram
TERRA AUTEM ERAT INANIS ET VACUA
ET TENEBRAE ERANT SUPER
  FACIEM ABISSI
ds ferebatur super aquas
dixitque ds fiat lux et facta est lux
et vidit ds lucem quod esset bona
et divisit lucem ac tenebras
appellavitque lucem diem
  et tenebras noctem
factumque est vespere
  et mane dies unus
dixit quoque ds
fiat firmamentum in medio aquarum
et dividat aquas ab aquis
et fecit ds firmamentum
divisitque aquas quae erant sub
  firmamento ab his quae erant
  super firmamentum
et factum est ita
vocavitque ds firmamentum
  caelum
et factum est vespere et mane
  dies secundus
dixit vero ds
congregentur aquae quae sub caelo
  sunt in locum unum
et appareat arida
factumque est ita
et vocavit ds aridam terram
congregationesque aquarum
  appellavit maria
et vidit ds quod esset bonum et ait
germinet terra herbam virentem
  et facientem semen
et lignum pomiferum faciens
  fructum iuxta genus suum
cuius semen in semet ipso sit super
  terram et factum est ita
et protulit terra herbam virentem
  et facientem semen
  iuxta genus suum
lignumque faciens fructum

et habens unumquodque sementem
  secundum speciem suam
et vidit ds quod esset bonum
et factum est vespere et mane
  dies tertius
dixit autem ds
fiant luminaria in firmamento caeli
et dividant diem ac noctem
et sint in signa et tempora
  et dies et annos
ut luceant in firmamento caeli
  et inluminent terram
et factum est ita
fecitque ds duo magna luminaria
luminare maius ut praeesset diei
et luminare minus ut praeesset
  nocti et stellas
et posuit eas in firmamento caeli
ut lucerent super terram
et praeessent diei ac nocti
et dividerent lucem ac tenebras
et vidit ds quod esset bonum
et factum est vespere et mane
  dies quartus
dixit etiam ds
producant aquae reptile
  animae viventis
et volatile super terram
  sub firmamento caeli
creavitque ds cete grandia
et omnem animam viventem
  atque motabilem
quam produxerant aquae
  in species suas
et omne volatile secundum
  genus suum
et vidit ds quod esset bonum
benedixitque eis dicens
crescite et multiplicamini
et replete aquas maris
avesque multiplicentur
  super terram
et factum est vespere
  et mane dies quintus

1 In principio creavit Deus cælum et terram.

2 Terra autem erat inanis et vacua, et tenebræ erant super faciem abyssi: et spiritus Dei ferebatur super aquas.

3 Dixitque Deus: Fiat lux. Et facta est lux.

4 Et vidit Deus lucem quod esset bona: et divisit lucem a tenebris.

5 Appellavitque lucem Diem, et tenebras Noctem: factumque est vespere et mane, dies unus.

6 Dixit quoque Deus: Fiat firmamentum in medio aquarum: et dividat aquas ab aquis.

7 Et fecit Deus firmamentum, divisitque aquas, quæ erant sub firmamento, ab his, quæ erant super firmamentum. Et factum est ita.

8 Vocavitque Deus firmamentum, Cælum: et factum est vespere et mane, dies secundus.

9 Dixit vero Deus: Congregentur aquæ, quæ sub cælo sunt, in locum unum: et appareat arida. Et factum est ita.

10 Et vocavit Deus aridam Terram, congregationesque aquarum appellavit Maria. Et vidit Deus quod esset bonum.

11 Et ait: Germinet terra herbam virentem, et facientem semen, et lignum pomiferum faciens fructum juxta genus suum, cujus semen in semetipso sit super terram. Et factum est ita.

12 Et protulit terra herbam virentem, et facientem semen juxta genus suum, lignumque faciens fructum, et habens unumquodque sementem secundum speciem suam. Et vidit Deus quod esset bonum.

13 Et factum est vespere et mane, dies tertius.

14 Dixit autem Deus: Fiant luminaria in firmamento cæli, et dividant diem ac noctem, et sint in signa et tempora, et dies et annos:

15 ut luceant in firmamento cæli, et illuminent terram. Et factum est ita.

16 Fecitque Deus duo luminaria magna: luminare majus, ut præesset diei: et luminare minus, ut præesset nocti: et stellas.

17 Et posuit eas in firmamento cæli, ut lucerent super terram,

18 et præessent diei ac nocti, et dividerent lucem ac tenebras. Et vidit Deus quod esset bonum.

19 Et factum est vespere et mane, dies quartus.

20 Dixit etiam Deus: Producant aquæ reptile animæ viventis, et volatile super terram sub firmamento cæli.

21 Creavitque Deus cete grandia, et omnem animam viventem atque motabilem, quam produxerant aquæ in species suas, et omne volatile secundum genus suum. Et vidit Deus quod esset bonum.

22 Benedixitque eis, dicens: Crescite, et multiplicamini, et replete aquas maris: avesque multiplicentur super terram.

23 Et factum est vespere et mane, dies quintus.

German Luther Bible

Das erste Buch Mose.

Das 1 Capitel.
Schöpfung der welt.

Im * anfang schuf GOtt † himmel und erde.   * Joh. 1, I. 3. Col. 1, 16. Ebr. 11, 3. † Pf. 33, 6. Pf. 107, 25.

2. Und die erde war wüste und leer, und es war finster auf der tiefe: und *der Geist Gottes schwebete auf dem wasser. Pf. 33, 6.

3. Und GOtt sprach: * Es werde licht. Und es ward licht.   * 2 Cor. 4, 6.

4. Und GOtt sahe, daß das licht gut war. Da * schied GOtt das licht von der finsterniß.   * Ef. 45, 7.

5. Und nante das licht tag, und die finsterniß nacht. Da ward aus abend und morgen der erste tag.

6. Und GOtt sprach: * Es werde eine veste zwischen den wassern; und die sey ein unterschied zwischen den wassern.   * Pf. 33, 6. Pf. 136, 5.

7. Da machte GOtt die veste, und schied * das wasser unter der veste, von dem wasser über der veste. Und es geschahe also.   * Pf. 104, 3. Pf. 148, 4. Jer. 10, 12. c. 51, 15.

8. Und GOtt nante die veste himmel. Da ward aus abend und morgen der andere tag.

9. Und GOtt sprach: Es samle sich * das wasser unter dem himmel an besondere örter, daß man das trockene sehe. Und es geschahe also.   * Hiob 38, 8. Pf. 33, 7. Pf. 104, 7. 9. Pf. 136, 6.

10. Und GOtt nante das trockene erde, und die samlung der wasser nante er meer. Und GOtt sahe, daß es gut war.

11. Und GOtt sprach: Es lasse die erde aufgehen gras und kraut, das sich besame; und fruchtbare bäume, da ein jeglicher nach seiner art frucht trage, und habe seinen eigenen samen bey sich selbst auf erden. Und es geschahe also.

12. Und die erde ließ aufgehen gras und kraut, das sich besamete, ein jegliches nach seiner art; und bäume, die da frucht trugen, und ihren eigenen samen bey sich selbst hatten, ein jeglicher nach seiner art. Und GOtt sahe, daß es gut war.

13. Da ward aus abend und morgen der dritte tag.

14. Und GOtt sprach: Es werden *lichter an der veste des himmels, die da scheiden tag und nacht, und geben zeichen, zeiten, tage und jahre, * Pf. 136, 7. Sir. 43, 2. 9.

15. Und seyn lichter an der veste des himmels, daß sie scheinen auf erden. Und es geschahe also.

16. Und GOtt machte zwey große lichter; ein großes licht, das * den tag regiere, und ein kleines licht, das die nacht regiere, dazu auch ‡sterne. * 5 M. 4, 19. ‡ Hiob 9, 9.

17. Und GOtt setzte sie an die veste des himmels, daß sie schienen auf die erde,

18. Und den tag und die nacht regiereten, und * schieden licht und finsterniß. Und GOtt sahe, daß es gut war. * Pf. 104, 20.

19. Da ward aus abend und morgen der vierte tag.

20. Und GOtt sprach: Es errege sich das wasser mit webenden und lebendigen thieren, und mit * gevögel, das auf erden unter der veste des himmels fliege. * c. 2, 19.

21. Und GOtt schuf große * walfische, und allerley thier, das da lebet und webet, und vom wasser erreget ward, ein jegliches nach seiner art: und allerley gefiedertes gevögel, ein jegliches nach seiner art. Und GOtt sahe, daß es gut war. * Pf. 104, 26. Hiob 40, 10.

22. Und GOtt segnete sie, und sprach: *Seyd fruchtbar und mehret euch, und erfüllet das wasser im meer; und das gevögel mehre sich auf erden. * v. 28. c. 8, 17. c. 9 2. 7.

A   23. Do

1 Am Anfang schuf Gott Himmel und Erde.

2 Und die Erde war wüst und leer, und es war finster auf der Tiefe; und der Geist Gottes schwebte auf dem Wasser.

3 Und Gott sprach: Es werde Licht! und es ward Licht.

4 Und Gott sah, daß das Licht gut war. Da schied Gott das Licht von der Finsternis

5 und nannte das Licht Tag und die Finsternis Nacht. Da ward aus Abend und Morgen der erste Tag.

6 Und Gott sprach: Es werde eine Feste zwischen den Wassern, und die sei ein Unterschied zwischen den Wassern.

7 Da machte Gott die Feste und schied das Wasser unter der Feste von dem Wasser über der Feste. Und es geschah also.

8 Und Gott nannte die Feste Himmel. Da ward aus Abend und Morgen der andere Tag.

9 Und Gott sprach: Es sammle sich das Wasser unter dem Himmel an besondere Örter, daß man das Trockene sehe. Und es geschah also.

10 Und Gott nannte das Trockene Erde, und die Sammlung der Wasser nannte er Meer. Und Gott sah, daß es gut war.

11 Und Gott sprach: Es lasse die Erde aufgehen Gras und Kraut, das sich besame, und fruchtbare Bäume, da ein jeglicher nach seiner Art Frucht trage und habe seinen eigenen Samen bei sich selbst auf Erden. Und es geschah also.

12 Und die Erde ließ aufgehen Gras und Kraut, das sich besamte, ein jegliches nach seiner Art, und Bäume, die da Frucht trugen und ihren eigenen Samen bei sich selbst hatten, ein jeglicher nach seiner Art. Und Gott sah, daß es gut war.

13 Da ward aus Abend und Morgen der dritte Tag.

14 Und Gott sprach: Es werden Lichter an der Feste des Himmels, die da scheiden Tag und Nacht und geben Zeichen, Zeiten, Tage und Jahre

15 und seien Lichter an der Feste des Himmels, daß sie scheinen auf Erden. Und es geschah also.

16 Und Gott machte zwei große Lichter: ein großes Licht, das den Tag regiere, und ein kleines Licht, das die Nacht regiere, dazu auch Sterne.

17 Und Gott setzte sie an die Feste des Himmels, daß sie schienen auf die Erde

18 und den Tag und die Nacht regierten und schieden Licht und Finsternis. Und Gott sah, daß es gut war.

19 Da ward aus Abend und Morgen der vierte Tag.

20 Und Gott sprach: Es errege sich das Wasser mit webenden und lebendigen Tieren, und Gevögel fliege auf Erden unter der Feste des Himmels.

21 Und Gott schuf große Walfische und allerlei Getier, daß da lebt und webt, davon das Wasser sich erregte, ein jegliches nach seiner Art, und allerlei gefiedertes Gevögel, ein jegliches nach seiner Art. Und Gott sah, daß es gut war.

22 Und Gott segnete sie und sprach: Seid fruchtbar und mehrt euch und erfüllt das Wasser im Meer; und das Gefieder mehre sich auf Erden.

Douay Rheims 1609

Creation. 1

THE BOOKE OF GENESIS, IN HEBREW BERESITH.

Chap. I.

God createth heauen and earth, and al things therin; distinguishing and beutifying the same; 26. last of al the sixth day he createth man: to whom he subiecteth al corporal things of this inferior world.

1 IN" THE beginning GOD created hea-
2 uen, and earth. † And the earth was
voide & vacant, and darkenes was
vpon the face of the deapth: and" the
Spirite of God moued ouer the wa-
3 ters. † And God said: Be light made.
4 And light was made. † And God
saw the light that it was good: & he
5 diuided the light from the darkenes. † And he called the
light, Day, and the darkenes, Night: and there was euening
6 & morning, that made one day. † God also said: Be :: a firmament made amidst the waters: and let it diuide betwene
7 waters & waters. † And God made a firmament, and diuided the waters, that were vnder the firmament, from those,
8 that were aboue the firmament. And it was so done. † And
God called the firmament, :: Heauen: and there was euening
9 & morning that made the second day. † God also said: Let
the waters that are vnder the heauen, be gathered together
into one place: and let the drie land appeare. And it was so
10 done. † And God called the drie land, Earth: and the gathering of waters together, he called Seas. And God sawe that
11 it was good. † And said: Let the earth shootforth grene
herbes, and such as may seede, & fruite trees yelding fruit
after his kinde, such as may haue seede in it selfe vpon the
12 earth. And it was so done. † And the earth brought forth

A grene

Act. 14, 15. 17, 24. Psalm. 32, 6. 135, 5. Eccli. 10, 1. Heb. 11, 3. Iob. 38. Ier. 10, 13.

The first part. Of the creatiō of al things.

The Church readeth this booke in her Office from Septuagesima til Passion Sunday.

Also this first chapter & beginning of the second on Easter Eue before Masse.

:: The firmament is al the space from the earth to the hieghest starres. the lowest part diuideth betwene the waters on the earth and the waters in the ayer. S. Aug. li. 11. de Gen. ad lit. c. 4

:: Likewise heauē is al the space aboue the earth. in whose lowest

1 In the beginning God created heaven, and earth.

2 And the earth was void and empty, and darkness was upon the face of the deep; and the spirit of God moved over the waters.

3 And God said: Be light made. And light was made.

4 And God saw the light that it was good; and he divided the light from the darkness.

5 And he called the light Day, and the dark-ness Night; and there was evening and morning one day.

6 And God said: Let there be a firmament made amidst the waters: and let it divide the waters from the waters.

7 And god made a firmament, and divided the waters that were under the firmament, from those that were above the firmament, and it was so.

8 And God called the firmament, Heaven; and the evening and morning were the second day.

9 God also said; Let the waters that are under the heaven, be gathered together into one place: and let the dry land appear. And it was so done.

10 And God called the dry land, Earth; and the gathering together of the waters, he called Seas. And God saw that it was good.

11 And he said: let the earth bring forth green herb, and such as may seed, and the fruit tree yielding fruit after its kind, which may have seed in itself upon the earth. And it was so done.

12 And the earth brought forth.

King James 1611

| The creation | Chap.j. | of the world. |

THE
FIRST BOOKE
OF MOSES,
called GENESIS.

CHAP. I.

1 The creation of Heauen and Earth, 3 of the light, 6 of the firmament, 9 of the earth separated from the waters, 11 and made fruitfull, 14 of the Sunne, Moone, and Starres, 20 of fish and fowle, 24 of beasts and cattell, 26 of Man in the Image of God. 29 Also the appointment of food.

IN the beginning God created the heauen, and the earth.

2 And the earth was without forme, and voyd, and darkenesse was vpon the face of the deepe: and the Spirit of God mooued vpon the face of the waters.

3 And God said, Let there be light: and there was light.

4 And God saw the light, that it was good: and God diuided the light from the darkenesse.

5 And God called the light, Day, and the darkenesse he called Night: and the euening and the morning were the first day.

6 And God said, Let there be a firmament in the midst of the waters: and let it diuide the waters from the waters.

7 And God made the firmament; and diuided the waters, which were vnder the firmament, from the waters, which were aboue the firmament: and it was so.

8 And God called the firmament, Heauen: and the euening and the morning were the second day.

9 And God said, Let the waters vnder the heauen be gathered together vnto one place, and let the dry land appeare: and it was so.

10 And God called the dry land, Earth, and the gathering together of the waters called hee, Seas: and God saw that it was good.

11 And God said, Let the Earth bring foorth grasse, the herbe yeelding seed, and the fruit tree, yeelding fruit after his kinde, whose seed is in it selfe, vpon the earth: and it was so.

12 And the earth brought foorth grasse, and herbe yeelding seed after his kinde, and the tree yeelding fruit, whose seed was in it selfe, after his kinde: and God saw that it was good.

13 And the euening and the morning were the third day.

14 And God said, Let there bee lights in the firmament of the heauen, to diuide the day from the night: and let them be for signes and for seasons, and for dayes and yeeres.

15 And let them be for lights in the firmament of the heauen, to giue light vpon the earth: and it was so.

16 And God made two great lights: the greater light to rule the day, and the lesser light to rule the night: he made the starres also.

17 And God set them in the firmament of the heauen, to giue light vpon the earth:

18 And to rule ouer the day, and ouer

1 In the beginning God created the Heauen, and the Earth.

2 And the earth was without forme, and voyd, and darkenesse was vpon the face of the deepe: and the Spirit of God mooued vpon the face of the waters.

3 And God said Let there be light: and there was light.

4 And God saw the light, that it was good: and God diuided the light from the darkenesse.

5 And God called the light, Day, and the darknesse he called Night: and the euening and the morning were the first day. 6 And God said, Let there be a firmament in the midst of the waters: and let it diuide the waters from the waters.

7 And God made the firmament; and diuided the waters, which were vnder the firmament, from the waters, which were aboue the firmament: and it was so.

8 And God called the firmament, Heauen: and the euening and the morning were the second day.

9 And God said, Let the waters vnder the heauen be gathered together vnto one place, and let the dry land appeare: and it was so.

10 And God called the drie land, Earth, and the gathering together of the waters called hee, Seas: and God saw that it was good.

11 And God said, Let the Earth bring foorth grasse, the herbe yeelding seed, and the fruit tree, yeelding fruit after his kinde, whose seed is in it selfe, vpon the earth: and it was so.

12 And the earth brought foorth grasse, and herbe yeelding seed after his kinde, and the tree yeelding fruit, whose seed was in it selfe, after his kinde: and God saw that it was good.

13 And the euening and the morning were the third day.

14 And God said, Let there bee lights in the firmament of the heauen, to diuide the day from the night: and let them be for signes and for seasons, and for dayes and yeeres.

15 And let them be for lights in the firmament of the heauen, to giue light vpon the earth: and it was so.

16 And God made two great lights: the greater light to rule the day, and the lesser light to rule the night: he made the starres also.

17 And God set them in the firmament of the heauen, to giue light vpon the earth: 18 And to rule ouer the day, and ouer the night, and to diuide the light from the darkenesse: and God saw that it was good

As you can see, the antiquated versions are similar and do not take the same level of liberties in translation that some modern versions do. It's a good research-practice to have copies of the pages of these older versions at your side when getting down to the deeper details of research. Using a term like "sky" in translation is just plain wrong. However, because some words don't really have a translatable word in other languages, the person translating will do their best to substitute a word, and as you will find as you read on, that can produce a very different interpretation by those of us who read the Bible's text today. The term "firmament" is not translatable and we must attempt to understand what it actually is rather than using the term "sky" in its place.

Using modern Bible versions that are accurate to the versions of antiquity just shown are the best to use as study Bibles. All other English modern versions are ultimately influenced by these versions either directly as the source of the modern translation, or indirectly because it was the version that the translator was familiar with.

Peering Deeper

When approaching this topic with an open mind, we have to first get a perspective on who was doing what in the text, and how exactly we interpret what we do interpret about the text from what we have already experienced or believe. For instance, what exactly is "Heaven" or the "Heavens"?

The word "Heaven" is like the word "Heave" or to lift up, but that is a very human perspective. Since everything in space from our Earthly perspective is up, if the word *Heaven* is derived from the word *heave* then it makes sense. But what if the word heave or even "heaven" are a base word, and we only think of it as up because to us it is up? Let that simmer in the back of your mind while you consider this next thought.

What is "earth"? Is it the ball of dirt on which we stand? Or is it the dirt that the ball is made of, such as when we excavate for a new building and use "earth" movers to take the earth or dirt from here to there? Or is it possible that it is even more fundamental than that? Could "earth" be the elements? Or even whatever the elements are made of?

Now re-read any version of the first sentence of Genesis that was previously shown mentioning "heaven(s)" and "earth" and think how different the text can then be interpreted. If "earth" in Genesis 1:1 is not our planet that we affectionately refer to as "Earth", then what is it? The possibility is very real here that we call Earth "Earth" not because it was created in the first sentence of Genesis, but rather it could be called so because it is made of the substance of "earth" that was being referred to in the first sentence of Genesis. The distinction here is important.

We also have to consider the potential influence on our minds of the term "the" being used, as in "In the beginning God created heaven and earth" Versus "In the beginning God created the heaven and the earth". These seemingly small differences can add up to a lot of influence in our thinking regarding our interpretation of all subsequent events in the Creation account.

Considering what was just proposed, we need to back up even before Genesis' first sentence and consider what was there before the so called "beginning", as in "In the beginning God Created..."

Before the Beginning - the Expanse of Space

Some Bible versions even go as far as translating it to be "In the beginning when God began creating", which also can give an entirely different view of what might have occurred. Understand that as of the writing of this book, there are many questions that we humans have yet to answer to a satisfactory point beyond all question and doubt. This is true not only about what occurred, but also about the various translations. So, I will propose other information, views, thoughts, and opinions not for you to act upon as advice, directive, recommendation, counsel, or any indication that you should follow what **I** say or think, but rather they are proposed for *you* to consider as possibilities that may be more, or possibly less, valid than many of the common theories you are likely familiar with today.

Some people think that everything was always just there, maybe not as we see it today, but that space debris abounded and collided and formed all of the celestial bodies that we see today.

Others think that there was nothing except a tiny point so small that it could not even be detected, and then suddenly it exploded in a big bang, eventually forming all that we see.

But both of those views avoid a vast emptiness with nothing at all in it, and both fail to explain how the single point or the space debris actually got there to begin with.

Then on the "believers" side, we have a Biblical perspective where God always existed and eventually created everything in six twenty-four-hour days.

There are also those who combine things a bit and think that God created it all with a big bang and thus they attribute everything in Creation to God's interaction in a grasping attempt to reconcile it with "science". It is worth noting here that Reverend Monsignor Georges Lemaitre is said to be the instigator of the very questionable "big bang" theory.

Now, realize here at this point that we are not asking about the Creation, but rather what existed **before** Creation, if anything existed at all. Did the elements exist and did God just use them to whip up the Universe? If you choose that particular belief then you are still stuck with explaining how the elements got there to begin with. We won't go any further down this particular path right now because it is a topic in itself.

For the purposes of this book we are going to be discussing pre-Creation as either empty space that is void of **all** things, versus the general scientific beliefs of either a big bang, or a static creation where base elements were just always there and formed everything we see.

If we take an approach that the base elements always existed, it leaves us with the question of: How did those elements get there to begin with? This is where the concept of a "God" helps to alleviate the need to explain that initial origin, which again is another topic.

A Fundamental Concept

Everything starts with an idea, and if there is no God then the big bang fills the void in our minds. But what if God actually is real? Assuming God is real, then the account of Creation in Genesis has some serious explaining to do in order for us to be able to reconcile it with reasonable scientific observations. "In the beginning God created heaven and earth." This statement in nearly all versions including those not listed, indicate that God created "heaven" and "earth". So, at the most base and fundamental concept we have to understand this as, these things were created and thus did not exist before Creation, according to the Bible.

What is Space?

The real question here is: Was "space" a part of Creation, or is space empty and has always existed as a vast emptiness of nothingness? What is space? Is "heaven" space? In regard to this question here in our case we are not referring to the Heavenly place where many people believe they will go when they die; but rather we are speaking only of the astrophysics view of "heaven" discussed in this book. Is *Space* the "heaven" or "heavens" mentioned in the very first sentence of the Bible? Did space exist before Creation? If it didn't, then what was there before? Was there more space?

Is it possible that the blackness of space is a created thing even though to us it is nothing? This particular part of the topic can be a bit esoteric and philosophical so we will leave the question open because it is somewhat incomprehensible to most people. You might be thinking on a Biblical basis that the "heaven" includes everything, and this may be so later on in the Creation text. However, the "everything" was *placed in* the "firmament" of the "heavens" when you read on in Genesis chapter One around verse 14 where God said "let there be lights made in the firmament of the heavens". It says "let there be lights made <u>in the</u> firmament of heaven" and that distinction is critical regarding this particular part of this subject. The lights were added to "heaven", so when considering that point, we can assume that "heaven" was initially empty regarding tangible visible created items.

This brings to mind the question of what the term "Firmament" means. The firmament is discussed later and is an integral term that must not be swapped out for a translator's best guess of meaning. Some things need to *not* be translated because they are meant to be understood as is. Firmament is one of those words that is the definition itself, and unless we *understand* what the word means, we cannot *know* what the word means. To grasp this you can consider the word "Truth". It is difficult to define without using the word itself or a derivation of it to define it. As you read on you will get a clearer understanding of "Firmament" if you do not already fully grasp it.

The Reality of
The Bible, Creator, Heaven, Earth, and Science

Here is the reality with the Bible, the Creator, heaven, earth, and science: If a Creator exists then we should be able to somehow prove it through some sort of found evidence. And the Bible may be one way to do this along with science when, and only when, we don't make stuff up. Speculation is fine to do, but when we insist that a theory is fact, then that is the moment when science goes down a dangerously erroneous path. We must proceed to do the analyses correctly and honestly. If there is a Creator who created all things, then the Bible and science should support that if either or both are true. But with regard to the Bible, there is each our own perception of particular events and the interpretation of our own perception—The same is true of our scientific findings. There is a great deal of evidence, but much of the evidence is often arbitrarily attributed to other imagined phenomena.

Since in our own modern times, we tend to view past people as "primitive", how could the ancient writer of Genesis One write the things that were written in it? It seems easy if it is written from an earthly perspective. But what if the perspective is not earthly, as was indirectly suggested throughout this entire chapter? What if the perspective is the Creator's perspective as told by a human? Then what does the first sentence of Genesis mean?

In that case, "heaven" in the very first sentence of Genesis One could be what we think of as space, or possibly even a non-tangible scientifically undetectable substance, and "earth" is the substance within that space or substance that all things are made from. If this proposal is at all correct then we must stop for a moment and realize that Genesis has insight beyond, both, our conventional understanding, and our desire to attempt to understand. What if heaven is a substance, and in this case a superstance?

Chapter 2

Is There a Creator God?

The most difficult question facing science is–Is there a Creator God? How do we prove something so intangible? If there is a God, we should somehow be able to prove it. For many Christians the answer is simple; "of course God exists!" After all, billions of people over thousands of years have been following the same "God"–that is to say the God of Abraham and Isaac, or for Muslims the God of Abraham and Ishmael. Plus, we have the Bible and the Quran.

But for science-minded people that's no evidence for a God at all. We should be able to find evidence, yet it appears to many that there is no evidence of a scientific nature to prove that God exists. Is this so? Is it true that there is no evidence whatsoever that God exists?

Illusions and Magicians

If you have ever been to a magic show, you may have seen the crowd amazed by the magician's assistant magically disappearing.

Or maybe somehow the card you picked from the deck of cards magically appears in your own pocket. How can this be? Most people understand that some sort of trick is involved, but there are those who are very amazed at what they just witnessed and simply do not understand what they saw. An experienced magician who is watching the very same magic show will immediately know exactly what just occurred in nearly any of the seemingly magic events performed, but they will be impressed at the *quality* of the skills of the performing magician regardless.

So what we have to ponder is, do believers see things as if it is some sort of hocus-pocus magic that God performs? I mean after all, "God said let there be" and then it suddenly came to be. Or at least that's what many people think they think about Creation. Are believers a bunch of unknowing suckers falling for the most obvious trick performed by the writers of the Bible's books and embellished by leaders of the various religions? Or, is it that people who say that they are scientific, believe that some Christians believe that "God said let there be" and then it suddenly came to be and that believers are a bunch of unknowing suckers falling for the most obvious trick performed by the writers of the Bible's books and embellished by leaders of the various religions?

Did I just repeat myself? Not exactly. This is the oldest switch trick in the book. There certainly are far too many believers that believe things in a blind-faith manner and will believe something because they *think* "the Bible says so" and/or because their preacher said so.

But there are also many believers who have been told what they believe by the non-believing so-called "science-minded" people. In that particular situation, the science-minded people believe that the Bible says that God created everything in six twenty-four-hour days, which simply does not reconcile with what we actually see in science. They then inadvertently or even deliberately pull the old switch trick on the unsuspecting believer and tell them that God could not have created everything in six twenty-four-hour days because of all of the scientific points they then will attempt to impose on the gullible listening believer. Then the unsuspecting believer, feeling that their God and their beliefs are being attacked, will dig in their heels in the debate and defend the position that was cast upon them by the "science-minded person"—This is a position that they never really fully thought out, but now feel compelled to defend nonetheless because their religion and overall beliefs are being attacked.

My question is this: Who is it here who really thinks the Bible says that everything was made in six twenty-four-hour days? In all three of the groups mentioned above, the most amazed group is the science-minded group. In fact, they are so perplexed and amazed by the tricks of Creation mentioned in the Bible that they will discount the Bible as fairy tales because the Earth was obviously not created in a single day nor was the Universe and everything in it created inside of a six twenty-four-hour-day timeframe. After all, science-minded people are too smart to believe such ridiculous tales.

The aforementioned blind-faith believers generally have little concern regarding the events of Creation and often will not take the bait when someone attempts to lure them into the debates surrounding the topic. In general, many people simply never give enough thought to seriously pose such questions in their own minds. There is nothing wrong with this position on personal spiritual basis. Just as one person cares nothing of ball-sports and another cares nothing of fishing, so also, some people care nothing of how Creation occurred. To them Creation occurred and they don't care why or how, because it's all just there. This is fine for that person, but will do little for their children as those children enter the world and then encounter and potentially be interested in the details surrounding the origins of everything.

Regarding the unwitting believers that allowed themselves to be lured into the debate when they were told that they believe that everything was created in six twenty-four-hour days, they are generally not technically/scientifically prepared to be in the debate at all. They are often more or less just trying to defend their God's honor and their own honor. It's not wise to fully engage in a debate for which you are not prepared, because you are certain to come across as a foolishly incorrect person even if you are correct.

The point that I am getting at here is that it is mostly the science-minded people who believe that the Bible says everything was made in six twenty-four-hour days, and so they disregard the Bible as if "it's just a bunch of stories". It is they who are most amazed by the Bible's Creation account; in fact, they are so amazed that they choose to not believe the Genesis account of Creation, or the Bible for that matter, because to them it does not agree with their scientific findings. This makes sense when we consider the reactions of people who are perplexed while watching a magic show as mentioned at the beginning of this section.

Science-minded people watch the magic show and when they see a trick, in their minds they say "Nah, this can't be, there must be some sort of slight-of-hand occurring here, the magician is somehow cheating. The rabbit didn't really disappear into thin air. It was shoved into a hidden compartment or something like that". Similarly, they look at the Bible's Creation account and understand that it took more than a day to create the "Earth" and so they proceed to explain as best they can without what to them are, the apparent tricks of the Bible.

But there is another group a bit more discerning than the three groups previously mentioned. This group ponders the Genesis account a bit more meticulously. Some of these people are believers and some are not. But both the believers and the non-believers in this group want nothing more than to find out what *actually* occurred, how it may have occurred, and if possible, why it occurred and maybe even, who did it. The two sectors within this more open-minded group are not necessarily in opposition to each other, even though the believers believe in

God and the non-believers claim they do not. During their quest over a period of time a few will switch beliefs because of what they think they have found, but generally they will still continue their quest to find the actual answers.

When people believe that the Bible is a hocus-pocus account of Creation, they discount it because they believe they have better explanations of Creation than the Bible gives. But just because we can explain why we are deceived when the assistant escapes through a secret door or into a compartment that is out of our viewing range, and just because we can explain how some of all of this works, does not mean that the magician is not there and it does not mean the he did not design and implement the equipment and techniques to produce the events you witnessed. If we choose to believe it is magic, and even if we are very wrong, it does not mean that the magician does not exist. What it means is that *we* do not understand how he did it, and *we* do not understand what all that he did means.

Remember these five groups because you yourself likely fall into one of them. The groups are:

- Blind faith believer
- Science-minded non-believer (Closed-Minded)
- Believer defending their own and their God's honor
- Open-minded believer (Science-minded)
- Open-minded non-believer (Science-minded)

Which group, if any, are you in?

God of the Gaps Perspective

The science-minded non-believer group will invoke the term "god of the gaps" when debating the topic with believers, because when something cannot be explained by natural phenomenon some believers will say "God did it" and then offer no further explanation. But just because someone can't explain things does not mean the magician doesn't exist or that the seemingly magical events did not occur at hand of the skilled magician. It means we don't understand *how* it happened and thus we cannot adequately explain it.

Uncertainty Principal

Science has a term called the "Uncertainty Principle". It roughly means that some things are difficult to measure. For instance, the *position* and *speed* of something small like an electron cannot be measured and viewed accurately at the same time because there are too many uncertainties and the measuring process itself can affect the measurements taken. I mention this only because science, as we think we know it today, is not as accurate as we want to imagine. Most things in astrophysics are best guesses or theories. The theories may be close to right, or even right, but they can also be, and often are, very inaccurate and outright wrong.

A *theory* is little more than an educated guess. It is typically what someone proposed in a scientific paper which was then distributed to their scientific peers for review. Those peers will then ponder, debate, and then try to test the theory to see if it holds up against scrutiny. If no one can disprove the theory with solid evidence, then the theory holds up, and often it will eventually be broadly accepted in the particular scientific discipline or area of study. But sometimes, after wide and long-term acceptance of a theory, new data comes to light that proves that the accepted theory, tested by and believed correct by peers, was wrong.

When we consider what "theories" actually are, and that science does indeed have an uncertainty within it, it is certainly scientifically valid to have a theory-of-God, meaning that someone can theorize that God does indeed exist if they have any evidence, or if their theory cannot be disproved.

Will Not Accept Any Evidence as Proof

If someone makes a statement that they "will not accept any evidence as proof of a Creator", then they truly are not open-minded and are unwilling to accept any truth that might be revealed regarding that possibility; this is truly an anti-science approach. In essence, they have decided that against any evidence shown they have chosen to not accept true facts, should those facts at any time be presented to them. It is quite unfortunate, but we hear far too many "science-minded" people make the very closed-minded statement that they "won't accept any evidence of a God". Of course, not all science-minded people say such foolish things, but the fact that *any* "science-minded" person would even consider that thought is clear proof that they are not open-minded at all, and they have chosen to be deliberately ignorant and scientifically blind.

It is Difficult to Prove Something that is Intangible

In science it's difficult to disprove a theory of something that is not tangible, such as God. This argument is often used by scientists and atheists to assert their claim of the absence of God. However, this applies to the big bang theory as well. It is difficult to disprove something that is not tangible when no evidence exists. Conspiracy theories also work this way when the alleged conspiracy never actually occurred.

We see this sort of thing in the salacious news reports of politicians and other famous people when the news people try to make the news interesting to their listeners and viewers. The news organizations only need to suggest mere impropriety by an elected official and then the official is forced into the insurmountable task of overcoming negative public opinion that was derived from the lies that were falsely reported to the people by the newscast.

In a scientific manner, we often view the issue of whether or not God exists as—prove that God exists then I will believe it. But the issue is more that if God does not exist then prove it. In other words, in science the view is: Your "Theory of God" does not exist. Then this is wrongly turned around and the burden of proof is then placed on the person who believes in God when there is already much evidence on their side, yet they don't realize this or don't see it clearly enough to properly articulate it to other people.

Science should apply the same standards to the issue of the non-existence of God as it does the big bang theory—It is more of a belief system than it is a theory. There are more anomalies in the theory of the big bang than most big bang believers will admit to and/or understand. A good lie will be built upon evidence that will make the lie appear true even though it is not true. When we build a sequence of small theories in order to build a larger theory, we are then trapped in utter speculation. For instance, the big bang theory is a broad theory that was once proposed, and then smaller theories were invented to support that broad theory. Those who believe in the theory will search until they exhaust any avenue and then they will start again on another avenue until they find a piece of evidence that *appears* to fit the broader theory, and then they proceed to fool themselves into believing that the broader theory is plausible and that it is based upon the sub-theories sought only because of the primary broader theory. This is the **incorrect** way to approach the subject.

Believing in the Bible is somewhat different with blind faith believers because the substantial amount of Bible information is handed to them, which they then attempt to explain by using the book itself as the evidence. This is also the **incorrect** way to approach the subject.

If the Creator Created, then evidence of that Creation should exist and we should not have to use the Bible as "proof" of Creation. The Bible is the theory, and what we find scientifically is the evidence.

The biggest mistake that those who believe that God Created everything make is that they often fail to be able to logically explain things found while doing scientific study, so they unnecessarily invent things to fill in those gaps or anomalies. Someone once foolishly said that God created the fossils to look old and that the fossils were created as fossils rather than occurring naturally. If there is a Creator that can make all things, then do we really need to imagine that the Creator needs to make things look old so that *we* can explain and justify the Bible? Not likely! This way of thinking is a backward approach and it is a very foolish one at that, and it is similar to big bang theory in rationale. If God did Create everything, then evidence should be everywhere.

Would a Creator Arbitrarily Dictate Rules or Laws

If you have ever read the Bible cover to cover you will find that there were many rules imposed on the people to assist in keeping them safe. These rules are often looked down upon in modern times, yet we all follow most of these rules with ease and should be grateful that we do, because they keep us healthy. Since the Genesis Creation account was given thousands of years ago before our technologically advanced microscopes and other such equipment, imagine trying to explain microbiology to people thousands of years ago who did not have access to a microscope and who possibly never imagined that such small things as microbiological entities exist.

Should we assume that the Creator Created evil for the sake of torturing us? Or can we take the more logical approach that evil is the risk of allowing us free will? To humanity's discredit, when given free will, there is a high number of us who will mess things up over and over again. If we are tempted by evil, is this just a mere mental thought. or is it perhaps more concrete than that? Are we being tempted by real life entities who will bring us to our own destruction?

Some of these things are discussed in a later volume of *The Science of God*, but the point being made here is that we don't always understand things as they were intended to be understood. And all too often, we foolishly invent explanations and then use those invented explanations as if they are the actual facts. This causes other people to discount what the person proposing the thought says or believes because of the many anomalies in their newly invented explanations.

Will we imagine that a Creator would arbitrarily dictate rules or laws and then impose them on us with fire and brimstone? Or should we be reasonable and realize that such events were likely warnings that were intended to advance us and possibly keep us safe? We have to stop placing *our meaning* into the words of others. Instead, we need take a more diligent approach to understand what was actually meant and then attempt to see things through the eyes of the writer.

If there is a God that Created, then the evidence should be overwhelming, and **that** God is not going to make things appear to be old to dupe us in order to satisfy the inadequacies in our own understanding. We all, both believers and science-minded atheists, need to pull our heads out of the sand and try to see things from perspectives other than our own narrow-minded non-provable beliefs.

Chapter 3

In the Mind of Man; What is a Day?

Considering everything discussed in the last chapter, you should be getting a pretty clear understanding of how we humans think and how unknowingly biased our scientific thoughts and conclusions typically are. Without understanding this aspect of humanity, we have little hope of ever grasping the realities of Creation regardless of how it occurred.

The mind of man is active and does wonder, but we have a difficult time freeing ourselves from our predispositions. And as a result, those predispositions influence what we choose to believe. Subsequently, what we believe then influences what we do and what we are willing to accept as evidence.

Are you open enough to view things from another perspective? Or will you remain stuck, as so many people are, in the many incorrect assumptions about the Bible and about actual science?

There is this incorrect thinking many of us have today where we imagine that people from the ancient days did not know things, or that they were stupid or ignorant. Too often we carry our nonsensical belief that things have only been known for the last two or three hundred years or so since the so-called "age of reason", yet nearly all evidence, both written and concrete, shows us otherwise.

Ancient Days

In Genesis One it says "In the beginning God created heaven and earth" and there is no reference to days. Regarding the text, we must not get too overly particular about periods and commas. This is because the grammar of the translator and also the translator's interpretation may have altered our modern perception, so we have to essentially look at the entire text without any punctuation.

Further, we must ignore capital and lower-case letters because in some of the ancient texts there doesn't seem to be much distinction in that regard, at least not to the extent that we see today. We must stop trying to *force* our own interpretation into the text. When removing all punctuation and capital letters from the first several verses of Genesis, the English Douay-Rheims version of the Bible reads as follows:

"in the beginning god created heaven and earth and the earth was void and empty and darkness was upon the face of the deep and the spirit of god moved over the waters and god said be light made and light was made and god saw the light that it was good and he divided the light from the darkness and he called the light day and the darkness night and there was evening and morning one day and god said let there be a firmament made amidst the waters and let it divide the waters from the waters and god made a firmament and divided the waters that were under the firmament from those that were above the firmament and it was so and god called the firmament heaven and the evening and morning were the second day..."

You will notice when reading Genesis One without punctuation or capital letters that your perception can no longer depend upon commas and periods, or on the importance of a word via capitalizations. Doing so forces you to have to read the text a bit differently and more carefully scrutinize every thought in the text and in your head.

As I have found over the years while having books edited, the various editors will disagree as to where a simple comma goes and will move commas that I myself may have placed in the text. To make things a bit more perplexing, during my own edit process I myself will move commas in one edit and then replace it in a subsequent edit of my own work. Commas are great tools, but they are a crutch that can help or, as often occurs, can cause problems when reading text. In legal writings, commas can become a point of contention in court proceedings regarding the specific meaning of the text. That's why legal text has long-winded sentences, often without commas or at least very few commas depending upon the essence of the legal document.

When we remove the punctuation, we are then forced to rethink the text and must group the text into subject matter. The text just shown from the first eight verses of Genesis One from the English Douay-Rheims Bible clearly has three distinct subject groups.

1) heaven and earth and waters
2) light and dark
3) firmament and waters

Are these aspects what we think they are? What are *heaven, earth, water, firmament, light,* and *dark*? And further, what are *face, deep, darkness, spirit, void,* and *empty*? And are *empty* and *void* the same thing?

Our problem, both scientifically and Biblically, is that in only the first eight verses out of many thousands of verses in the Bible, we generally do not properly understand many of the words used. If we or the Bible are to be taken seriously, we have to work to understand, first and foremost, the perspective of the person instructing. And second, we must work to understand the meaning of the words that *they* used, and then finally the context in which those words were used.

There is phonetics such as *Their, There,* and *They're* where they all sound the same, but have very different meanings and then there is the issue of a single word such as *Read* pronounced *Reed,* or *Read* pronounced *Red*—pretense or past tense. We also have single words with multiple meanings such as the *bark* of a tree or a dog's *bark.*

So, I ask you, does "*day*" in the case of Creation really mean a twenty-four-hour day?

If the first several verses are broken into subject groups, then the first "day" could not have occurred until the first day was completed. In that case "heaven and earth" were created first and then God said "light be made" and it was the first day. This would make any event that occurred before light was made to have been timeless. Meaning that there would have been no way to track time. It is highly unlikely that the term "day" as used in the first couple of sentences of Genesis was *ever* meant to mean a single twenty-four-hour Earth day.

It was only light that was Created and it was not night and day as we perceive it today from our spinning Earth. Some people might get upset at that thought and think that I am debating the Bible, but that is simply not so. Here is the text again but with more verses that follow:

"in the beginning god created heaven and earth and the earth was void and empty and darkness was upon the face of the deep and the spirit of god moved over the waters and god said be light made and light was made and god saw the light that it was good and he divided the light from the darkness and he called the light day and the darkness night and there was evening and morning one day and god said let there be a firmament made amidst the waters and let it divide the waters from the waters and god made a firmament and divided the waters that were under the firmament from those that were above the firmament and it was so and god called the firmament heaven and the evening and morning were the second day god also said let the waters that are under the heaven be gathered together into one place and let the dry land appear and it was so done and god called the dry land earth and the gathering

together of the waters he called seas and god saw that it was good and he said
let the earth bring forth the green herb and such as may seed and the fruit tree
yielding fruit after its kind which may have seed in itself upon the earth and it
was so done and the earth brought forth the green herb and such as yieldeth
seed according to its kind and the tree that beareth fruit having seed each one
according to its kind and god saw that it was good and the evening and the
morning were the third day and god said let there be lights made in the
firmament of heaven to divide the day and the night and let them be for signs
and for seasons and for days and years to shine in the firmament of heaven
and to give light upon the earth and it was so done and god made two great
lights a greater light to rule the day and a lesser light to rule the night and the
stars and he set them in the firmament of heaven to shine upon the earth and
to rule the day and the night and to divide the light and the darkness and god
saw that it was good and the evening and morning were the fourth day"

As you will see if you read the text carefully, after light was
Created, the light and dark where divided or separated. We tend
to read this as "night and day" as we understand it today, but it is
not possible that our current idea of night and day are what was
being referred to here. If you read toward the end of the
additional verses you will notice that "god made two great lights a
greater light to rule the day and a lesser light to rule the night and the stars
and he set them in the firmament of heaven to shine upon the earth and to
rule the day and the night and to divide the light and the darkness"

Here we need to ask if this is nothing more than a rehashing
of the first two sentences of Genesis, or is this a specific event
separate and apart from the events of the first sentence? This
will depend upon what "earth" of the first sentence actually is.

"in the beginning god created heaven and earth and the earth was void and
empty and darkness was upon the face of the deep and the spirit of god moved
over the waters and god said be light made and light was made"

Then later in the text:

"let the dry land appear and it was so done and god called the dry land earth"

This seems to be a very different picture than the first
sentence. We can be reasonably certain that the first sentence is
a group of events in itself and that the "heaven" and "earth" was
made and then light was made and divided from the dark. It was
the phenomenon that we call "light" with photons and such that

was likely created here, because "god created heaven and earth and the earth was void and empty and darkness was upon the face of the deep". "earth" was already there when the light was made to be.

And then

"god said be light made and light was made"

And then later

"god also said let the waters that are under the heaven be gathered together into one place and let the dry land appear and it was so done and god called the dry land earth"

Here something was named to be "Earth" *after* the light was made and *after* "the waters that are under the heaven be gathered together into one place and let the dry land appear"

And then

"let there be lights made in the firmament of heaven to divide the day and the night and let them be for signs and for seasons and for days and years to shine in the firmament of heaven and to give light upon the earth and it was so done and god made two great lights"

The order of events in the text are critical in working to scientifically analyze the text. And the meaning of *heaven, earth, darkness,* and *waters* in the first couple of sentences are critical for our understanding of Genesis and of Creation. If you think on all of this for a bit and re-read the text with this in mind, then you should see quite clearly that "Earth", as it is named in the text, is made of dry land or somehow separated from the "waters" and the other "earth" substance had already existed before that point in the text. Then the lights were made that would shine upon the Earth.

The importance of this is crucial because if you view this all from our modern understanding of *light, day, night, earth,* and *water* then you will not be able to see the order of Creation in a scientific way and you will tend to gloss over the text likely thinking that that our Sun was made on the first day and that our

Earth was already there on day one–but it was not! We will be examining some of these issues more later on.

The Philosophers

Ancient philosophers pondered these sorts of questions. But the Greek philosophers did not have the Bible as we have it compiled today, nor did they have our modern science. They likely had portions of the Bible, or certain books that are included in our modern Bibles, but the book we so affectionately call "The Bible" as it is today, is a collection of books that was held by the leaders of the people. The books that were more reliably documentable were eventually canonized as the book of the people, which is to say "The Bible". You can read more about this in the book *Understanding The Bible - The Bible How-To Manual AND The Things We Don't See.*

Many of the Greek philosophers lived a couple of hundred years BC. People of that time would not have the extreme Biblical biases and influences which we have today. They would still have had to deal with the same human nature as we do and many of the same questions we have, but they did *not* have television or the internet, or books to the extent that we use as guides today. In that ancient time, the general populous was hungry for information and only the thinkers and those who studied had such information to offer, thus giving an impression of great wisdom; and in most cases this was true in relative terms because the thinkers back then likely had access to far more information than the general populous had access to.

As human as we are today, I feel confident that human nature was very similar thousands of years back. The point being that just as we today have our daily trials and troubles and have little time to ponder these things, so too would this have been the case thousands of years back. History records this to be so, and so it appears that no matter what period of history we examine, human troubles abound and it is only by each our own choice that we can work to remove these troubles from our own lives.

The ancients likely had a bit of an advantage over us today because some of these people had the privilege of being the first person ever to think the thoughts, meaning that they did not have the biases that have been created over the years since then to influence their thinking. What we now have in that regard can be both a blessing and a hindrance to us, so we must use care in our research and analysis of findings.

The Pharisees

If you have read the New Testament of the Bible you will recall that The Christ was not particularly fond of the attitude of some of the Scribes and Pharisees. It was probably not all of them that Christ was upset with, but rather those who imposed their own version of morality onto the people when the Pharisees themselves would not even live up to their own version of morality.

There were debates regarding—the soul ending at death, versus having eternal life—and also whether or not there is a hell and a heaven. These debates occurred between the Pharisees, the Sadducees, and the Essenes.

Many of what we would think of as *the religious people* from back then were what we might consider "scientists" today as well as being teachers and even priests and writers in our modern era.

In the same way that the Pharisees of Christ's days were referred to as hypocrites, the same is true in our modern era with too many people of science and of the church regarding Creation. Many of today's scientists are today's Scribes and Pharisees. Scribes and Pharisees are basically the writers and the speakers in our modern culture. Papers are written about scientific issues and if you don't agree with them, then "shame on you!" If you question this then you need only look at the debate regarding so-called "global warming" or the more versatile term "climate change." If you disagree with the premise of it, you will be viciously defamed by them and attacked in public forums.

The attitude is that their way is the only way and their interpretation of the evidence is the only possible interpretation. There is far more error in science than in religion in this regard, but that is more due to the fact that the religious believers often won't enter into the discussion about such topics.

Science is a great field of study that is being perverted by the Scribes and Pharisees of today. Let us not follow science Scribes and Pharisees or religious Scribes and Pharisees to our own destruction, because some of their many numbers simply do not add up.

An Expert's Opinion

For years we have listened to "experts" opine their ambiguous thoughts to the masses with me believing that somehow, they were "in the loop" and that I was missing out and not understanding what they said. But as was later found, many of their kingdoms of lies and inaccuracies cannot be understood because we cannot understand incorrect information. We can understand that it is wrong, but it is technically impossible to understand *incorrect* information.

I have read books and watched videos, too many to recall them all, and one disturbing trend in many of them is what was just alluded to in the previous sections. People *claim* to be "experts", but much of what they demand to be "fact" and "true" is actually nothing more than mere speculation filled with tremendous error. It is important to understand that there is nothing particularly wrong with speculation containing errors in the speculation. The problem comes in when it is demanded that we all accept that the erred speculation is a "fact". What this is getting at is, don't pretend to be the expert while you have many faulty conclusions. Instead, be the expert because you know your topic well and what you say has a solid foundation and holds up well against *all* scrutiny, and further, that it is admittedly open to modification when better information or discoveries are provided. We do not need to have the answers to everything in order to help the thoughts of each other move to a new state of understanding, we just need to be accurate and admit when something is speculation, versus it being actual provable fact.

In God We Trust, the Text

Getting back to "heaven" and "earth", let's examine the first three groupings:

1)   heaven and earth and waters
2)   light and dark
3)   firmament and waters

heaven, earth, water, firmament, light, and dark
but also:
face, deep, darkness, spirit, void, and empty

These are all key words that we need to begin to properly understand in order to understand Biblical Creation in unison with science.

You should by now have a fairly solid foundation in understanding that "heaven" and "earth" in the first sentence of Genesis One are not our sky and not the Earth we walk upon. And the "waters" are not the water we splash around and swim in. And the "light" is not our star the Sun. But rather our Earth was called earth because it is made of "earth" substance and water is called Water because it is a "water" substance and shares the fluidity attribute. The question we need to attempt to answer is what are those substances?

When students of any age go off to school, their parents are essentially saying to the children that the teacher is okay and is trusted. Then the student understands this as, whatever the teacher says is accurate. But often what the teacher says is *not* accurate when it comes to these types of topics.

Students who are believers typically go off to college and fall into full belief with big bang because the preachers of such theories have many seemingly convincing arguments that "prove" that the Bible is wrong and is full of incorrect information. Yet, when you test their data and their arguments you will find that they too interpret the Bible inaccurately just like they do their own data about their own theories. And those inaccurate interpretations are then projected upon the students who have not been given any tools whatsoever with which to make their own wise decisions and intelligibly discuss topics such as actual Creation.

The troubling part about this teaching issue is that the students end up never hearing the sort of discussions that a book such as this will invoke. And due to the humiliation that is cast upon them, typically by an ignorant teacher or professor, when the student attempt's to enter a debate or defend a position for which they are not prepared, they ultimately cower in defeat and reject the faith of their youth, thus abandoning any possibility of learning the actual truths about Creation.

Chapter 4

Finding Our Way

You have probably noticed by this point, that this book is more than just a book about the science behind Creation and astrophysics. We have a serious problem in our modern culture that is becoming increasingly worse. Too few people with incorrect information have been affecting too many unsuspecting people. All of society must collectively find our way, in an honest and open manner, if we ever want to rise to new levels of scientific understanding.

Someone could take some of my statements as ridiculous because of the missions to Mars and all that humans have been doing with rockets over the decades. But that's exactly my point. We have been playing with rockets for decades, and other than being a bit more efficient and having better computers allowing us to more quickly and more accurately achieve our goals, those rockets are never going to get us anywhere near another solar system in any of our lifetimes with our current rationale.

We have problems understanding the idea behind simple things such as "Black Holes" and "Singularity" because of the abuses of mathematics and sensationalizing of science for pop-culture notoriety.

When we fail to grasp the *basic* concepts behind the words used in science, we then end up having problems understanding causing us to invent meanings because we are trying to get it all to make sense to ourselves. And you can add to that that we do the same thing with the entire Bible. The problem we have regarding the Bible and science is that all too often we believe that somehow these two areas of study are separate and apart and that there is no, or should be no, connection whatsoever between them. But this simply is not the case and is not true in any way. In fact, it is quite the opposite. The Bible and science **must** agree on nearly everything or either the Bible or science or possibly both are utter frauds.

The term God is fine, but in the context that we are using within in this book, "God" is not a proper term. We must use the term "Creator", because in this book the fundamental topic is "Creation". And when discussing the Bible in relation to science with regard to Creation, the two, that is to say the Bible and science, are **inseparable**. The only thing that separates them is the misunderstanding within our own minds regarding this topic. Too many people, both Biblical and scientific, insist that the Bible says Creation occurred in six twenty-four-hour days, but science shows us otherwise.

However, if you review the last chapter of this book and when breaking the first four days down a bit, you can begin to see that the idea of six twenty-four-hour days is highly unlikely because the text itself indicates otherwise when read from the perspective of the Creator. In fact, it indicates otherwise from an Earthly perspective as well, that is unless you take liberties with the text that are illogical based upon the actual order of events as described within the text.

Our Curiosity is a Gift Abused

Curiosity is a gift that we abuse with our biases, but our biases come from our own blindnesses. Earlier a bit about the importance of the meanings of words was mentioned. In small-talk during general conversation, our words typically don't much matter as long as we are polite and kind to one another. But when discussing technical specifics in science or any other detailed field of study, then the words used and the meanings that we assign to those words are considerably more critical and can cause substantial differences in outcome and/or substantial disagreements in regard to each our own understanding.

Science is Not Owned by the Pop Culture Scientists

Notoriety is not a bad thing unless the people who admire those persons of notoriety *blindly* follow them. "Science" is a broad term that covers many areas of research. Basically, anyone who studies the deeper inner workings of things can be considered a scientist, yet we mostly associate the term "scientist" with those who study physics and astrophysics, which includes the study of the Earth and everything within it.

When it comes to the sciences surrounding the questions encountered in the Creation topics, there are many more views than have so far been presented in the previous chapters of this book. Those views are also a part of the points being made throughout this book.

Not all "believers" see the account of Creation as a six twenty-four-hour day event. There are many who realize that it was longer and who further believe that the "days" mentioned in Genesis are arbitrary time frames. And then there are also those who believe that it was six twenty-four-hour days, but theorize before the first day there was a very long undetermined time period of chaos.

We also have those who are heavily involved in science but also are committed believers who, in their own minds, understand that things may have taken somewhat longer than the Bible *appears* to indicate. Yet somehow, they are able to disconnect the two disciplines even though personally they see many similarities between the two disciplines.

Then of course there are the atheists who claim to believe nothing with regard to a Creator. And while it may be true that some people hold no opinion in this regard, if you do not believe in God or in a discerning Creator, then your belief is that this so-called "Creator" does not exist, and that assertion of non-existence is, in fact, a belief system.

What we all need to understand is that we all believe that which we believe, and those beliefs affect our scientific rationale. If we stand stubbornly by while trying to "prove" our own interpretation of the Bible to be true and that there is a God, then we will overlook good evidence that doesn't fit our beliefs when it is presented to us. This is true even if that evidence already supports what we ultimately want to believe.

For instance, if someone is very firm in their choice and has chosen to believe that everything occurred in six twenty-four-hour days, then they may feel that this book appears as if it is in opposition to what they want to believe. Where on the other hand, if a person is an ardent atheist and also is a scientist whose mind is made up then they too might feel that this book is in opposition to what they believe.

A belief is a belief and has no basis in fact. However, it is possible that our belief actually does match what is factual. We can see fact after fact after fact and still not believe what we have just seen. It seems silly that this could ever occur but it happens daily where people will disregard facts and continue in an erred belief system, even after seeing the actual facts. This is a blind faith issue which is typically attributed to "believers" alone, but the truth of the matter is that blind faith abounds in science even more pronounced than in religion.

It is saddening to watch young adults try to argue the so-called "facts" that they heard from their mentors who they are only repeating without discerning for themselves the religious or scientific talking points that they regurgitate.

If you understand a topic well, you will welcome opposition in discussion and you won't have any need to shout the opposition down or feel the need to shut them up. This is because if anyone listening has an open heart and an open mind they will see the errors in someone's erred proposals when those erred proposals are forced to stand next to accurate proposals–Truth reveals error. However, when someone has inaccurate beliefs and happens to be involved in such discussions, then what typically occurs when the facts begin to go against their erred belief system is that they will dig in to defend their non-defendable position in order to save face. This sort of erred self-defense makes themselves look evermore foolish. They will mock and ridicule in effort to hide their shame, digging an ever-deeper hole of folly for themselves to wallow in. This is true of both believers and of non-believing science people who follow erred "information", or better stated as **un**formation.

Other Religions

How many religions are there? There is no answer to that question because the number keeps changing. Take for instance the Christian religions. There was a point in time were Christians were all what we call today "Catholics" or the *universal* religion who believed that Jesus the Christ was the Promised One as discussed in the book *Understanding The Church - Upon This Rock I Will Build My Church*. Then along came the "Reformation" causing fractures in the beliefs of *Universals*, that is to say *Catholics*. Then each faction broke into smaller factions, and so it continues to this day causing much contention regarding topics of morality and many scientific aspects of the Bible. But divisions in beliefs have truly been occurring for thousands of years, such as the differences in opinions of the Scribes, Sadducees, Pharisees, and Essenes mentioned earlier.

Of the many religions that we know of today, most of them, even those that are ***non***-Christian, still generally believe that there is some sort of ultimate "God" above any other gods or other venerated entities. But the debates that we are familiar with in Western-world culture are generally between the Christian view of Creation and the scientific view of creation. Of course, the pop-science idea of "creation" is a misnomer, if there is no God then a "**creation**" cannot exist because there is no creator to create, so in many science circles the mysterious big bang is the source to all things being here.

For the purposes of this book we are speaking generally about *religion*, meaning–those who adhere to the teaching and discipline of the Biblical and Quaranical Abraham and his God the Creator. That is so say "the God of Abraham, Isaac, and Jacob for the Christians and Jews", but "the God of Abraham and Ishmael" for the Muslims. Ishmael is the son of Abraham through Sarah's handmaid Hagar, and Isaac is the son of Sarah herself who is Abraham's wife.

Question the Bible

In many ways it is good that people question the Bible because questioning is how we understand what is truth and what is not truth. If we fail to question things then we are stuck in whatever belief we have chosen, even if it is not correct. If there is a God, then it is highly *unlikely* that that God would want us to be wrong and believe wrong things. In fact, if you read the Bible, you should have noticed that was somewhat a point of contention between God and the people. The people would follow folly, and their doing so was upsetting to God because the people would repeatedly choose to follow a path filled with error.

I once heard someone actually say that we should doubt everything when it comes to science, but doubting everything is a fool's approach and is a biased approach. If someone asks another person if he or she believes in God, and then the other person says "Nah, I doubt it" then it is clear they have a sort of bias in their thinking, otherwise they would have said "I am uncertain as to whether or not a God exists."

We need not doubt things, but we should question things, especially if someone has information that conflicts with our current assessment of the information that we have personally gleaned from the entire world of data.

Doubting is okay provided that our agenda does not override the truth. The flawed logic to discredit the Bible that is used scientifically, today, is unparalleled in history. Believing that the Sun rises and sets is better logic than much of the logic that is used to discredit the Bible in our modern era. Why is this you ask? It is because when the Sun rises in the morning we all witness it nearly every day and we watch it traverse across the sky until it finally rests on, and then plunges below, the opposing horizon. To a mind that refuses to inquire as to how the Sun gets back to the eastern horizon to start again, all sorts of speculations could be made, such as a *new* sun coming up every day.

Everything in the Bible points to an understanding of circles and movement along the concept of a circle. But because we have a prejudicial understanding of what we think they meant when someone wrote the words found in the Bible, we wrongly assume that they believed that the world was flat. In reality it is only those who use the same type of flawed logic that is used to deduce the big bang who believe that many people thought the world was flat and that the Sun travelled around the Earth, in past times. It is likely true that some people did, but more likely that many did not believe the Sun travelled around the Earth.

Some Translations of the Bible

Septuagint
Vulgate
Masoretic
Guttenberg Vulgate
JPS Tanakh 1917
Webster's Bible Translation
English Revised Version
King James Bible
American King James Version
King James 2000 Bible
New American Standard Bible
Holman Christian Standard Bible
NET Bible
New Living Translation
Jubilee Bible 2000
American Standard Version
Darby Bible Translation
World English Bible
New American Standard 1977
GOD'S WORD Translation
Young's Literal Translation
English Standard Version
International Standard Version
And the list goes on...

If you were to go back to the oldest, most ancient copies of the writings we now see in the Bible and you were to learn the language and then try to translate those texts; your interpretation would still be influenced by your current understanding of the text through what you have already experienced. Thus if any errors exist in current translations then you are very likely to make the same errors in your translation unless you only translate verbatim. But even translating verbatim is somewhat subjective because at some point there will be more than one choice of meaning for a particular word to be translated to. However, with all of this in consideration, we must also credit the Bible for being very consistent when we consider how many renditions have been offered over the many years and the relative consistency with which many modern versions concur with the ancient Biblical documents.

Of course there are other documents that conflict with the Bible as well, but we are speaking about the Bible's translation and its consistency. Even the more obscure newer modern language translations tend to drive the main theme points and basic facts fairly accurately. Now, as for how many generations of translations there are, that is also often overblown. Some people believe that the Bible had been translated and retranslated over and over since it has been assembled into what we today call "The Bible". The actual quantity of generations of translations is fairly small as discussed in the book *Understanding The Bible - The Bible How-To Manual AND The Things We Don't See*, because every so often someone or some group comes along and says "what does it really say?" And "what does it really mean?" They then will strive to understand the oldest held, most ancient texts and proceed to translate those into their modern language so that they are sure that it is being translated to their own satisfaction. Now sure, we can have some discrepancies, but the details that are in agreement are too numerous to have **all** translations wrong. And in all of the translations, the major points of contention still exist, which are: Is there a God? Is there Heaven and Hell? And, is there a "Christ"? If there is a God then does any of it matter? And, how was the Earth formed or Created? These points are very important regarding the Bible.

Now, as for translation, versus honest translation, if someone was very agenda-driven they might proceed to correct what they perceive as errors or discrepancies in the Bible, yet many of those perceived "errors" are left as-is in other translations, which is why it is so easy to point out these *perceived* discrepancies that are altered in some versions. This speaks for the accuracy of the translation of the text over the centuries. When an older text is again translated directly into a modern language it remains largely consistent with the translations of the translations. This is due to the deep dedication for accuracy that the translators had with regard to the process because they feel obligated to be correct in their work rather than only translating what they themselves perceived it to mean. Yet we cannot discount the potential errors from the translation process. These points are incredibly important to understand in this discussion about Creation.

There are modern era authors who complained of inter-language translations of their own books during the era of time when printing became abundant in the seventeen-hundreds. Thus, it is reasoned that if their own works were so full of errors through a single translation to another language, the Bible would then even be more so because it has been translated from previous translations. While that is a good point, it is almost altogether overridden by the fact that the translators of the Bible were dedicated to accuracy. And they seldom gained much from the translation process, which was a labor of love rather than of money. They were instructed to translate it accurately which is why the many translations today are largely in agreement. Everyone is just trying to figure it out. The translation issue is partly why there was resistance to translating the Bible to other languages to begin with. Parallel translations by other translators help to confirm the accuracy of each translation. You can find more thoughts in this in the book *Understanding The Bible - The Bible How-To Manual AND The Things We Don't See.*

Human Direction-Whose Bible Should We Use

The debates surrounding the most reliable versions of the Bible will have a different level of importance depending upon how detailed the scrutiny of a particular subject is. The Living Bible says "14 Then God said, "Let lights appear in the sky to separate the day from the night." Notice the use of the term "sky". This is a tremendous overreach in the text. In the Book of "Job" the term "bulldozer" is used in one Bible. Such paraphrased translations are easy to read and could in some cases be okay for someone who never read a Bible before, but these types of translations are very bad to use if you ever intend on doing research or scientific study with them and eventually enter in to debate on such topics.

The best versions are those that are closest to the oldest source documents available. The Latin Vulgate is a good Bible to use, but must be translated to our native language in order for us to read it. The Catholic Bible in the past generally used the Douay-Rheims English translation, which was initially translated from the Latin Vulgate and later influenced by the King James version and has been quite consistent over the years and can be easily compared to past printings of antiquity. Many other versions are derived from it or from its source.

"Firmament" is the term that is most logical when discussing the astrophysics aspects of Creation, as opposed to the term "sky". This is especially true with regard specifically to exactly when the "firmament" comes into the picture. Various Bible versions use terms like "expanse", "canopy", "vault", "sky", or even "horizon". Such terms are found in the more recent late post-Reformation translations, where on the other hand, earlier more trustworthy Reformation translations all use the term "firmament" even in their early foreign tongue. This could be thought of as a trivial point of semantics, but it is far more critical than that when working to fully understanding the Genesis One text, as you will see as you read on.

"Canopy", "vault", "sky", and "horizon" imply a very different picture and will instantly cause any reader who is new to the text to have an erred interpretation of the text. They will most certainly derive that the text indicates an Earth-centric view. Not necessarily that they will assume that our Sun revolves around the Earth, but rather in their view of Genesis' first four days, they will see it all as pertaining specifically to our planet Earth. This perspective will cause all sorts of anomalies in their thinking and interpretation of the text by these new readers, thus causing them to view the division of waters as the clouds and the seas, with the sky in between. Such erred views will cause many to eventually leave their unsubstantiated beliefs behind, along with the rest of the Bible and their religion as all too often happens.

The term "expanse" is only slightly better as it takes us beyond our Earth's atmosphere. Yet it is significantly different than "firmament". The term "expanse" evokes a sense of something expanding like a big bang. The term "expanse" is used in the come-lately Bible's such as *World English Bible, New American Standard Bible, English Standard Version, Young's Literal Translation* and others. This brings us to a point discussed earlier about which Bible version to use during deeper Creation scrutiny.

The influence of the translation is obvious. When the first translation using "expanse" came about, then other Bible translations followed. Is this accurate? Not likely. The use of the term "expanse" came after the expansion theories arose before expansion became commonly known as "the big bang". The idea of expansion as previously mentioned is not a particularly new idea. And the understanding of "expanse" versus "firmament" is very different depending upon what is actually occurring on the second day of Creation.

"and god said let there be a firmament made amidst the waters and let it divide the waters from the waters and god made a firmament and divided the waters that were under the firmament from those that were above the firmament and it was so and god called the firmament heaven and the evening and morning were the second day".

The more authoritative term, "firmament", implies a *firming* or a *making* action in the text, but the term "expanse" implies a division of distance. Then there are more seemingly insignificant points to consider. For instance, the phrase "in the midst" implies between two distant points of the expanse. "Expanse" however, implies an entirely different view than is being explained using the term "firmament" in this volume of *The Science of God.*

What we need to look at is the oldest available or most original text's actual words used. For instance, "amidst" as is used in the Douay version can imply "in with" or "within" or "together" where the phrase "in the midst of" implies being in between things. These differences seem small but greatly impact the scope of view and the possible interpretation of the Genesis Creation account.

Genesis is a bit light on details regarding Creation, but it is even worse when using some modern versions of the Bible, thus making it impossible to come to realistic scientific conclusions. Translations are difficult to do because the translator can only use the understanding that they have of each word to best convert it to the new language. The secondary step for a language-to-language translator is to try to take the words that don't make sense to them, and then in context apply their best guess of a word or description from the new language of translation. But doing so typically compromises the text a bit because sometimes there simply is no word to adequately translate to in the new language, and in that case the word should be left as is.

There are no other choices when translating the text of any book. The problem that occurs with language-to-language translations is that the translator will influence the text that is being translated into the new language with *their own understanding* of the events described. There is no other way to translate. Their understanding will reflect the culture that is contemporary to them at that time, almost as a photo-like capture of the times. This means that as we humans attempt to figure out how everything came to be, and we propose our theories to the world, if those theories take root in the minds of the people, then translations written of anything that has any relation in subject matter with the theory will be influenced by the newly rooted theory. Additionally, our view of their understanding and past culture will affect our interpretation of what we, today, think they believed.

There is nothing wrong with this when the theories are correct, but what if the theories are in error in any way? The answer is that those errors are then indirectly written into the translation of connected subjects. To get a better view of this problem we have to take a simple English-to-English translation of the Bible where the person translating is doing it for modern understanding and readability. While working to remove such words as the archaic *thee, thine, doeth* etc... the modern language translator will replace them with *you, your,* and *does* etc. This is typically not an issue in understanding when reading the modernized version, but when translation liberties are taken with words, such as "firmament" and translating it to "horizon" or "sky" then that presents a real problem in true understanding, especially with regard to the scientific aspects of the Genesis text.

But fear not, if you use the Douay-Rheims version with all of its 73 books rather than a 66 book modern post-reformation Bibles, you won't spontaneously burst into ball of flames and suddenly become a Catholic. The two Bible types are nearly identical aside of the other books stripped from the recent post-reformation protestant Bible versions.

Latin and English Douay-Rheims versions are among the clearest stated, however it is good to look at other versions when very particular hair-splitting detail is required. Generally, stay clear of paraphrased versions due to the liberty that has been taken with regard to translated word choices.

The best way to state this is that when you study, use a Douay or Catholic version of Genesis or even a King James Version and the older they are the better. If you can read or work through Greek, try to get images or facsimiles of some of the very old Septuagint text or at minimum an older Latin Vulgate translation.

Newcomers to Bible study will make the mistake of getting a copy of a Modern "Hebrew" Bible and imagine it to be authoritative. But in reality, it holds the same and sometimes even less equivalence depending upon the version, than do many modern versions derived from the Ancient Hebrew, Greek Septuagint, and/or the Latin Vulgate, because some and even many modern Hebrew Bible versions are translated from the more recent Bibles rather than the from ancient Hebrew or

Greek. Translating the text from Greek and Hebrew and then into Latin and English and then back into "Hebrew" only adds another layer of confusion. But this is not true of all "Hebrew" translations, as some are from the ninth century Masoretic text, which however has an ambiguous origin. The point is that you need to select your study source-material with care, because some modern Bible printings have been modified to a point of unacceptable interpretations when being used for study purposes, which is not true of most versions of antiquity. If you happen to be so fortunate to find a treasure containing the actual originally written Hebrew version, then that would be an entirely different situation, because that is the ultimate goal of any Bible scholar.

The Douay Version and the King James Version of the Bible are among the most reliably translated. And when questions arise, use Douay, King James, and even Masoretic together to determine which is likely the most accurate to actual findings in science.

It really depends upon what you are doing with a Bible in regards to the version you might use. Modern versions are fine for a general overview of Biblical events, but when you are getting down to the deeper details, such as the science of Creation, especially if you are going to debate the topic, then it's a bit different. So seeking the oldest texts you can get a hold of is a wise choice. The older the authentic source is, then the less likely they are to have multiple layers of interpretation error.

So which version should we use? Use one with the oldest authority, and then use the various older versions when the toughest questions arise. Then balance the facts with the way each are stated in order to find which is likely the most accurate to science, if any. However, we must realize that translations that occurred after the theory of evolution and the big bang theory were imposed onto the people, will most likely have been influenced by those theories, so it is not a good idea to use them for objective study because they will often support those theories as they are biased on those subjects, and will *not* help you arrive at truth.

Stellar Directions

As humans who grew up in an established world, we don't have the same advantage of neutral thinking that Adam and Eve would have had. Everything we perceive, we perceive through our already formed subconscious view of what we see and hear based upon what we have already seen and heard and experienced. So for us, Earth is the ball of dirt that we stand upon and north is up on most of our maps. And when facing that direction, we then have east to our right hand and west to our left hand and south to our backs.

In later parts of the Bible, cardinal directions of North, South, East, and West are at times referred to in a ways that seem to not be relative to land or Earth surface directions and may be beyond our earthly directions. Is it possible that there is a cardinal direction in space or heaven? Again, in this book we are referring to the "heaven" of astrophysics and not a Heaven for souls.

Could north, south, east, and west be stellar directions rather than earthly directions? If so, could they also be used in a navigational way such as we do on Earth? When reading something we tend to always read it from an Earth perspective, causing us to lose the ability to see it from a stellar perspective. Is north really north as we think of it here on Earth? Or is north a direction in space that never changes? When you head north on Earth, then once you pass the Earth's North Pole you are then heading south, and again when you are heading south and pass the south pole you are then once again heading north. But with east and west you always keep going in the same cardinal direction. This point about stellar and cardinal directions is just another point to consider when trying to grasp the large amount of information contained in the Bible and in science.

If you were in space and were far enough away from our Sun, then the Big Dipper we are all familiar with that we see to our Earthly north would no longer look like a Big Dipper and you would likely be lost. Our sky, or heaven, and the apparent position of the stars in that sky, are based only upon our earthly perspective view. That view would quickly change if you could travel light years of distance from our Earthly home and look to the same general area of space where the Big Dipper resides. Our Earthly view of Creation is very shortsighted.

Chapter 5

The Spirit Moved

The text reads "and the spirit of god moved over the waters". Can we make any scientific conclusions regarding the "Spirit" based upon the text of Genesis One? Since there is only one reference to "Spirit" in the Creation account in Genesis One, we have little to compare against, making it difficult to draw any scientific conclusions. In context you can see that making any "scientific" conclusions would be reaching a bit far. Read in context:

"in the beginning god created heaven and earth and the earth was void and empty and darkness was upon the face of the deep <u>and the **spirit** of god moved over the waters</u> and god said be light made and light was made and god saw the light that it was good and he divided the light from the darkness and he called the light day and the darkness night and there was evening and morning one day"

When we have nothing to compare against, then it is difficult to determine much scientific significance of a word using that single word only once when it has no other contextual bearing that we can detect. Depending upon the English language version, the next time the term "Spirit" is used is in the sixth chapter of Genesis long after the Creation account occurred as follows:

"and god said: my spirit shall not remain in man forever, because he is flesh, and his days shall be a hundred and twenty years."

But in the Latin version of the Vulgate Bible in the second chapter of Genesis where it gives more details about man being made it reads:

Douay English Genesis 2:7 "And the Lord God formed man of the slime of the earth: and breathed into his face the breath of life, and man became a living soul."

Douay Latin Genesis 2:7 "Formavit igitur Dominus Deus hominem de limo terrae, et inspiravit in faciem ejus spiraculum vitae, et factus est homo in animam viventem."

Douay English Genesis 1:2 "And the earth was void and empty, and darkness was upon the face of the deep; and the spirit of God moved over the waters."

Douay Latin Genesis 1:2 "Terra autem erat inanis et vacua, et tenebrae erant super faciem abyssi : et spiritus Dei ferebatur super aquas."

The indications we have of Spirit in the Bible, references wind or breath or air and is associated with silence or quiet. Since our assumptions of the Creator are that the Creator is not concrete as we humans are, and the Creator has no physical parts as we humans do, we are forced into a somewhat narrow set of choices as to what *spirit* specifically is, yet we really don't know scientifically.

Is the Creator Spirit or did the Creator Create Spirit? In the Bible it doesn't get in to vivid detail about what specifically the Spirit is. Are we to assume that the references to Spirit are for the most part all referencing the Spirit of the Creator? In my own reading of the Bible I have little choice but to gather that the "*Spirit*" spoken of throughout most of the Bible is only one aspect of the Creator. And further, that aspect was possibly Created by the Creator, but is of a different nature than are the concrete parts of Creation. That's because the Spirit is spoken of as an entity with an ability to discern.

I am still seeking answers to the question of Spirit and its role in Creation. It appears that the Spirit of God was possibly surveilling the tangible nature of Creation and maybe wondering upon it and analyzing it, and then Created Light.

When I speak of the Concrete nature of things I am referring to the parts that form what we call the elements of our periodic table of elements, which are the building blocks of the parts of Creation that we visually see with our eyes today in outer space, or even the tangible aspect of our own hands.

The Spirit of Science

The Spirit of Science... What is it? In our modern world we often think of "Spirit" in terms of attitude as in the "spirit of '76" or "team spirit". And in that regard the spirit of science is nothing more than just an attitude or mood about science.

But let's take another look at the text:

"in the beginning god created heaven and earth and the earth was void and empty and darkness was upon the face of the deep <u>and the spirit of god moved over the waters</u> and god said be light made and light was made and god saw the light that it was good"

It says "and the spirit moved", and then it says "and God said let light be made". This appears to be an indication that the Spirit was somehow interactive in the discernment or analysis of the physical parts of Creation which at that point had already been in the works. If this is not the case then there is no significance in it even being in the text.

The "Spirit of God" is the ultimate scientist, or at least is a part or aspect of the ultimate scientist, and is the source of all science and reason as we understand science and reason today. It is interesting to listen to anyone who discusses how and why we humans are so very different from the animals. Sure there are similarities to the animals, but the differences are quite obvious. Yet, we can't quite put our finger on exactly what it is that makes us different or why it is so. That subject is discussed in more detail in *The Science Of God Volume 4 - Day Six - Evolution versus Man - In Our Image.*

The Spirit that hovered is likely the very same spirit that was placed into man, yet many of us disregard the event that is likely the single thing that separates us from the animals.

If there really is a God and that God is the Creator who made all things as described in Genesis One, then the brief documentation of that Creation as told in Genesis One should be taken seriously by all of science. Not to be followed and obeyed, but rather to be scrutinized and understood.

We must understand that the words that would have been in the very first inscribing of Genesis One would likely have all had a very specific purpose for being there. They would have been conveying the most important event that has ever occurred since before the counting of time began. It is highly unlikely that those words would have been used recklessly. But, we have to wade through, at minimum, two or three translations to try to understand the spirit of the text.

Some words in the Creation text such as "spirit" are very consistently used in nearly all credible versions of the Bible that are used for serious study.

The "Spirit of God" is possibly the very first scientist, but is in itself not necessarily the entirety of the Creator.

Blind Perceptions

If we are to take the Bible serious at all, then we have to decide if we are going to take the text as it is or if we will throw out parts of the Old Testament's Genesis One because they don't make any sense to us. If we are going to disregard any parts of Genesis One then there is no purpose in looking at any of it at all.

If the words or derivations of those words are there in a consistent manner in all of the credible versions of the Bible used for study, then we have to regard them as critical to the text. However, the problem we run into as humans is our uncompromising predispositions that are borne in our often erred beliefs. This causes us to have to cheat the words in effort to make those words match *our own erred understanding* of the text, rather than changing our understanding to match the text's actual intent.

We tend to view things through the filter of our limited experience, and that perception of the world is our blind-spot. We incorrectly make the assumption that long ago things were then, as they are today. We also assume that *our own perception* of the so-called "facts" we found from the past are true and accurate as to how we see those past events through today's eyes.

Consider the issue of radiometric dating. Our ability to test using carbon dating or any other radiometric dating methods makes certain assumptions. We can confidently say "assumptions" because of one simple fact: our versions of radiometric dating are limited to our experience and we can only make *extrapolations* as to the meaning of the atomic decay or activity. This means that we are assuming that what we see right now today is the same as it was thousands or millions of years ago. But this misses several factors that will not be able to be proven for billions of years. I feel safe in saying that few to none of the seasoned scientists and ardent supporters of these dating methods will be here to make the case for the accuracy of these current perceptions even eighty years from now. Thus, to assume perfection in our perceptions of one billion or more years back in time is simply non-provable to our current experience.

I am not suggesting that it is altogether wrong or that it has zero credibility, but rather, that radiometric dating is only a tool of speculation. People have dated newly created rocks from recent volcanic activity that occurred during our lifetimes, such as from the Mt St Helens eruption in 1980, and have been returned results of tens of thousands of years old when they knew it was only a couple of decades because they witnessed it all with their very own eyes.

Some of the dating methods could potentially be very accurate but we could be misreading the results or misunderstanding the results, or we could be reading the results with an inaccurate scale. When theorizing we must be careful to be able to separate the vast speculations from the actual reality.

Is the Bible Short-Sighted?

For some people, the current perception of the Bible is that it is short-sighted and did not imagine the heavens to be as vast as we now are beginning to realize when we utilize our orbiting telescopes, yet the Bible spoke quite broadly about the "heavens". Take a look are the term "the deep" in context and in two languages:

Douay English Genesis 1:2 "And the earth was void and empty, and darkness was upon the face of the <u>deep</u> and the spirit of God moved over the waters"

Douay Latin Genesis 1:2 "Terra autem erat inanis et vacua, et tenebrae erant super faciem <u>abyssi</u> et spiritus Dei ferebatur super aquas"

Notice use of the terms the "Abyssi" and "Deep". The term "Abyssi" is our word for "abyss". We often think of abyss as deep or bottomless, but since space has no direction we can quite safely infer that it was likely referring to *endless*. In other words, there is no known end to space. This again points back to the short-sightedness of our own life experience. If we look into a deep hole on Earth we are often somewhat awestruck. If you doubt this then consider the many annual visitors to the relatively shallow Grand Canyon. In the Grand Canyon we can actually see the bottom, and yet we are still awestruck by its vast size along with its depth and beauty.

What Meaning was Intended?

Our perception of religion and science are often not what was intended by the promoters of certain thoughts over the past hundreds or even thousands of years regarding those disciplines.

Often an outspoken person will be disgusted with their contemporaries in the Church or in the various scientific fields, so they speak out against the perverted belief systems that their contemporaries are pushing onto the public. When these outspoken people rise up against their contemporary counterparts they usually make statements against the misinterpretations that their counterparts are promoting. These often strong statements are then eventually understood by people decades or centuries or even millennia later. In such cases, people with an agenda that disregards a God will take the words of these outspoken people and pervert those words to mean something that the words were never intended to mean.

If someone speaks against a *religion* then does it mean that that person does not believe in a singular and supreme Creator? No, it does not. But this is how many people will misinterpret distaste for a religion. And in truth it is not distaste for a religion, rather it is distaste for other people's interpretation of that religion. It is the inaccurate and biased interpretation of things that we so deeply dislike.

Someone can say things correctly, but because of how *we* have come to understand the world and religions incorrectly, we then inadvertently pervert the words of those who correctly relay the information to us.

There are many who do this very thing with Albert Einstein's work. When Einstein said that "God does not roll dice", people will take his comments about God to mean that he did not believe in God at all. And all too often people mistake Einstein for an atheist because he was both a scientist and a Jew. Sadly some people mistakenly apply the "atheist" label to people who don't believe that Christ was the Messiah. But while Einstein, the Jew, might not have accepted Christ to be the Messiah, he did believe in the same God that Christians do.

Those who pervert Einstein's words in that way are usually outspoken atheists or agnostics and they want their idol, Mr. Einstein, to have the same beliefs regarding religion and God as they have. Thus, their beliefs are projected onto him. A person who believes in a Creator will typically not understand Einstein's comments in the same way as an atheist might understand them. Our perception of other people's words is largely based upon our understanding and perception of life, the world, religion, and science.

This is all the same as referenced earlier with regard to the meaning of the words used in Genesis. We all have agendas and some of those agendas include our desire to match the facts with our personal beliefs. But for others who are truly honest, their main desire is to match their own understanding to the actual facts and truth of what is said in the Bible and what it truly intended to indicate and to what actually exists in reality.

Solar Flares and Time Perception

As you have likely noticed "perception" is a primary topic in this book. We *must* deal with our biases, with the biases of people whose information we consider, with biases of scientific findings, and the biases from our perception of all of it. Of course, those biases cover a broad range of life, but we are focusing more on the specific biases that affect our perception of what occurred during Creation and how long it may have taken.

One of the more specific perceptions that we need to address and at least attempt to understand is that of "Time Perception". What I am about to say is quite simple, but often is difficult for people to understand and sometimes even to accept because of our chosen belief of how things work.

Our perception of time is typically in *seconds, minute, hours, days, weeks, months, years, centuries* etc. However, every one of those is only a compounding of the previous at 60 seconds in a minute and 60 minutes in an hour and 24 hours in a day and so on. But all of these are nothing more than our attempt to break up the time that it takes for the Earth to rotate one full turn which is 360 degrees. But, while we can compound 365 days into a year, a year really is not 365 days as we think of it. A year is really the duration it takes our Earth to travel around our star, the Sun. This path around the Sun starts at a very specific position of Earth's axial tilt perpendicular to the Sun and in relation to the various stars that surround us, and then finally to return to that very specific location in space relative to the Sun and other stars. A day on the other hand is specifically a single rotation of our globe, the Earth. This is an important distinction to understand scientifically. It may not have much significance in terms of billions of years, but it is an important distinction nonetheless.

There is a belief that our clocks of today are more accurate than ever before. Take for instance an atomic clock with the ability to measure very fine slices of time. Atomic clocks are very precise as they divide the time into very fine slices, so they could not possibly be inaccurate, right? Wrong! This is completely wrong thinking. The *resolution* of a clock has nothing to do with its *accuracy*. If we adjust that clock *off* from the normal setting then it is no longer accurate, yet it can still granulate the time to very small fractions. *Accuracy* of a clock as we think of time in relation to the rotation of Earth is what we are really trying to achieve.

We have satellites in space that orbit the Earth, and when solar flares occur, sometimes the clocks need to be adjusted so that they are accurate again. Those adjustments are indeed typically very small, and for all practical purposes they are undetectable without another atomic clock to compare with.

What I am pointing out here is that our perception of time through the use of time-keeping devices can be affected by the

Sun. And if a solar flare can affect our best "most accurate" clocks, then what can we assume would occur over billions of years regarding that perception? These clocks are said to be affected by gravity as well, and the further they are from Earth then to more they deviate from clock times here on Earth.

Some people could try to foolishly assert through this information that clocks got so messed up that six twenty-four-hour days could actually be realistic regarding creation. But since in the grand scheme of Creation, *mechanical* time-keeping clocks are a relatively new invention in human history, so that point is utterly irrelevant and mechanical clocks are certainly no better than atomic clocks.

Our time-keeping devices are handy for sure, but they have little relevance with regard to measuring the elapsed time preceding the conclusion of Creation.

Chapter 6

The Gravity of the Situation

It has been said many times by astrophysicists with regard to the big bang that during the point of singularity, gravity did not exist. They conclude therefore that the big bang could rapidly increase in size from its single point size of zero, indicating that *gravity* had been holding it as it sat in its alleged state of singularity.

Re-examine the first paragraph of this chapter. If singularity mathematically requires gravity, and gravity did not exist at the point of big bang explosion as some claim, then we have an anomaly in the rationale for that theory. If you study into this avenue of the theory you will find the point being made here to be accurate about those statements. While we can expect that some mystical theory will be invented to overcome this big bang anomaly, we must mentally cheat or ignore the anomaly for a split second in order for the math to be able to explain the big bang.

If you pay close attention, you will often hear people say or theorize that "the laws of physics didn't exist before big bang", and that those laws only came in to existence *after* the bang was initiated. But even at that, the big bang stretches the imagination further than it stretches the heavens. Gravity is the fundamental point of the laws of physics. It cannot hold or bring anything into singularity if it doesn't exist. If we cannot see through these ridiculous anomalies in our thinking, then we are doomed to repeat our folly, causing scientific progress to be slow, if not altogether non-existent.

Considering what we do when using our mathematics in the big bang theory, it is quite easy to see how narrowly we choose to interpret the words that are assigned to the Creation events that are written in the Bible. There is a more detailed account of some of the points about the big bang anomalies in *Bending the Ruler – Time Travel, The Speed of Light, Gravity, and The Big Bang.*

The Gravity of the Situation

Some of points made here could come across as confusing because they are often pointing out the errors found in the theories and interpretations of what we often hear and read about the topic of how it all came to be. The point being that it is difficult to explain an error because an error technically doesn't really exist; therefore it technically is *not understandable*.

The big bang relies on the mathematics used to calculate the gravitational pull of a particular mass. So while we can "mathematically" calculate something, it does not mean that it was possible to occur in reality. This does not mean that something is not possible, but rather that it is not certain and/or absolute just because the "math works". Some mathematicians say that everything is built with mathematics, and there are some, who are believers in God who have stated that "God used math" to build the Universe. But while this may be true it is probably not an accurate perception.

As nearly any prototype designer-builder will tell you, when you have a vision of what you want and you then build it, you must deal with the results and modify accordingly. Sometimes math is a part of that and sometimes it just doesn't matter. This is more likely how Creation occurred–likely as an ongoing effort to make something useful and amazing. And then when it was all done, due to its vast repetition and consistency we can now use our invented mathematics to predict gravity's effects.

Here's a really basic example of mathematic extrapolation that just won't work in reality. If you were to hold a simple table spoon in your hand that is filled with dirt from our planet Earth and you were able to stand on the Sun, the spoon would bend from the weight of the dirt due to the extra gravitational pull from the mass of the Sun, in theory. We can calculate this using math and we have some experience with varying gravity from going to the moon and experiencing its low gravitational pull. But, besides instantly being vaporized due to the intense heat of the Sun, you wouldn't be able to stand on the Sun because you also would be equally heavier on the Sun just as the dirt in the spoon was. You would be crushed by your own weight. The point is, just because we can calculate it using math doesn't mean the theories we are promoting through the use of that math could actually occur in a real physical sense.

Making Things Firm

Gravity is one of the most important points to be understood in the Bible's Genesis One account of Creation. Is gravity Created? If so, was it Created in the accounts stated in Genesis One? Could it have been Created later on? Or did it always exist and God just used it to do the job of Creation? Or is it a result of the Created? The importance of understanding this particular point is unparalleled in the entire discussion about how everything came to be and the Bible's accounts of Creation. Gravity is the key point of big bang, but seems to not be specifically referenced in Genesis One as "gravity". So how do we deal with this? To start, let's take a look at the first four days of the Creation text again with some key points underlined:

"in the beginning god created heaven and earth and the earth was void and empty and darkness was upon the face of the deep and the spirit of god moved over the waters and god said be <u>light</u> made and <u>light</u> was made and god saw the <u>light</u> that it was good and he <u>divided</u> the <u>light</u> from the <u>darkness</u> and he called the <u>light</u> day and the darkness night and there was evening and morning one day and god said let there be a <u>firmament</u> made <u>amidst</u> the <u>waters</u> and let it <u>divide</u> the <u>waters</u> from the <u>waters</u> and god made a <u>firmament</u> and <u>divided</u> the <u>waters</u> that were <u>under</u> the <u>firmament</u> from those that were <u>above</u> the <u>firmament</u> and it was so and god called the <u>firmament</u> <u>heaven</u> and the evening and morning were the second day god also said let the waters that are under the heaven be gathered together into one place and let the dry land appear and it was so done and god called the dry land earth and the gathering together of the waters he called seas and god saw that it was good and he said let the earth bring forth the green herb and such as may seed and the fruit tree yielding fruit after its kind which may have seed in itself upon the earth and it was so done and the earth brought forth the green herb and such as yieldeth seed according to its kind and the tree that beareth fruit having seed each one according to its kind and god saw that it was good and the evening and the morning were the third day and god said let there be lights made in the firmament of heaven to divide the day and the night and let them be for signs and for seasons and for days and years to shine in the firmament of heaven and to give light upon the earth and it was so done and god made two great lights a greater light to rule the day and a lesser light to rule the night and the stars and he set them in the firmament of heaven to shine upon the earth and to rule the day and the night and to divide the light and the darkness and god saw that it was good and the evening and morning were the fourth day"

You will notice several underlined words: *light, divided, darkness, firmament, amidst, waters, under, above,* and *heaven.* Some of these words are location or directional and also one became a name, but others we still have not come fully to understand.

We typically use the words *amidst, under,* and *above* as location or direction. The firmament was named "Heaven" as in "God called the firmament Heaven". "Heaven" being named in verse eight is often understood as the same "heaven" as listed in verse one, but this cannot be because the light had to be somewhere first and the text actually reads as "heaven" and "earth" were the first created aspects of Creation as far as we can tell. It is within that "heaven" that light was Created.

Then on the second day the "firmament" was created and was specifically named "Heaven". Is any of this important? I suppose that will all depend upon what we choose to interpret these words to mean. I would imagine that most of the people in this world would understand the idea of the term "firm" as something that is stiff or tangible that is able to be handled in some manner. The word "firmament" is obviously either derived from "firm or firm is derived from "firmament" If we think of "firmament" as to make firm, or assist in making firm, or the cause of making firm then the "firmament" begins to make sense from a scientific and physics perspective. Can we think of the "firmament" as gravity? And what really is gravity?

Gravity is our current best description of what is occurring rather than what it actually is. The root of the word "gravity" is "grave", like "engrave" or "dig a grave". In German it is "grab". Often in phonetic language migrations, "b" and "v" are interchangeable. Think of engraving or digging a grave as grabbing or pulling out the material. It is the grabbing or pulling that we associate with our understanding of gravity. You could think of it is Grab-ity.

I fully expect some people to resist that idea that the "firmament" is what we think of as gravity for a few reasons, such as, it is named "Heaven" as in the heavens where the stars reside. Yet we don't actually scientifically understand what gravity is–at all.

Light being made prior to gravity would make sense since the phenomenon of light itself is not caused by gravity nor does it depend upon gravity to exist as far as we know, even though it is mildly submissive to gravity.

The point this is getting at is that the phenomenon that we call "gravity" likely exists *because of* the firmament rather than being the firmament. So, while gravity might not have had to have been specifically Created, it is nonetheless a likely result of the firmament and therefore did not exist at least until the point of firmament being Created. This is not a perfect example, but rain is rain and you getting wet is a result of that rain hitting you. Gravity is also a result, caused by something.

Reading the Periodic Table

In our modern understanding of physics, we have the Periodic Table of Elements. We're not going to get into the dirty details of all of the elements here, but rather a quick look at the basics.

Atoms are "elements" that make up everything currently known to mankind, as far as we can tell. Some atoms have a different quantity of each of three parts. The three parts are *Protons*, *Neutrons*, and *Electrons*. The combination of the varying quantities of these three parts are what makes the difference between the helium in a balloon and a solid pure bar of gold at Fort Knox, or a pure iron bar, or even a piece of charcoal.

Things like gold, iron, helium, plutonium, or the even oxygen in the air that we breathe are all basic elements that we find on the Periodic Table of Elements. The interesting part about this is that the weight of a large pure cluster with an equal count of any element on a scale will vary depending upon the count of the three parts making up the specific element. Understand that we are referring to billions or trillions of any specific element in the clusters being weighed.

This ingenious system of Creation consists of three basic parts that when combined in given quantities will reliably become a specific element. This is true to a magnitude unlike anything we humans have been able to cause by ourselves with all of our making of things. Three simple pieces combined in various quantities makes all of the known elements. And then those elements in various combinations make everything we are and see and hear and taste and smell and touch or feel. Regardless of how it all occurred, it is really quite astonishing!

Predictions versus Speculations

With astonishing accuracy, we can take the figures from the periodic table and make many predictions. We, as a world of human scientists, will proudly show our prowess in this regard. It is true that it is impressive that we are able to do what we do; however, what is astonishing is the utter and absolute consistency that three types of parts that construct any atom in specific combination are so consistent that we humans are able to look at the size of a planet, for instance, and calculate how fast we must approach a planet so as to not surpass it, but instead actually get caught in an orbit around it. Then we are able to calculate just how quickly we will descend our interplanetary vehicle and are actually able to land it on a planet so far from us with such accuracy. This is only possible because of the unparalleled consistency of the elements and the parts thereof.

The margin for error in any human endeavor is really massive in comparison with the elements. If any of the three basic components of an atom did not have perfect, or at least near to perfect consistency, then there would potentially be massive differences when using those figures to calculate the gravitational pull of a particular planet.

Sir Isaac Newton was able to make predictions based upon his mathematical formulas. His fundamental formulas have held up for a very long time and are used by nearly all astrophysicists. When we base something on sound mathematics, then we are able to predict how long, how far, how much, etc.

But when we theorize, we are often combining prediction with speculations. A speculation is more of a guess, rather than using specific calculations in a prediction. Please do not misunderstand this to be in anyway opposed to proposing a theory, because it is not. We must propose our theories at some point and then set out to see if those theories hold up against honest scientific scrutiny. Anyone can make up far-reaching or false theories to sensationalizing aspects of science for the purpose of notoriety, but those erred theories will always eventually come to ruin along with those who refuse to release the erred theories from their mind. Let us all work to build sound theories, but then accept when those theories are proven to be incorrect.

Chapter 7

Dividing Our Mental Boxes

We tend to simplify things in our minds to a level that we find personally comprehendible to us as individuals, such as assuming that a hydrogen atom somehow formed and then when many of those formed they drew together until they were in such mass that they heated up and through that heat and pressure, nuclear fusion occurred compounding them into more complex elements. In our minds many believe this scientific theory is what caused the atoms to form more complex atoms and through this, *all* of the elements we know today have been created.

I am not specifically rejecting this theory. However, if a simple hydrogen atom can form of its own accord, then why not a more complex atom, with, let's say, two electrons or four? This is not to say that such theories are wrong, but rather that we simply do not know and are merely speculating on what we see as "logical" based upon our current knowledge and chosen beliefs. If we accept the idea that the more complex atoms formed from fusion of hydrogen or helium atoms, then we have to first ask how those hydrogen or helium atoms formed to begin with.

The problem we face with this is that our knowledge is not really "knowledge", but rather, it is all best guesses. Some of our guesses are provable to an extent, but many guesses simply are not provable as of the point in time when this book was written. Just as there are more possible routes to get from one location to another when you drive in a car, so too, is it possible that atoms could be created in ways other than what we have invented and insisted in our minds "scientifically". But, even proving that simple atoms can form more complex atoms through "nucleosynthesis" from fusion does not prove that this is how all complex atoms formed.

When we're looking in the wrong room to find something, we will never find the thing we seek in that wrong room. Thus, sometimes we must abandon our search and start anew. When we start again with a fresh perspective, it is amazing what we can accomplish, discover, and test.

We build our mental boxes in this same way when we decide that we understand something fully, when in actuality we are blinded by our preconceptions to a point where we fail to look elsewhere, thus disregarding and therefore never checking in any other mental box for our answers. This most notably occurs in science and in religion, but we also see it in areas such as politics.

Is Creation Misinterpreted?

Creation is amongst the most misinterpreted of events that man has ever known. This is mostly because of our Earth-life-experience. From the moment we are born, we are told how many "days" old we are, and as we age that turns to how many "years" old we are. Then, as we get older, we read the Bible in a way where we choose to understand it as, on the first "day" this was made and on the second "day" that was made. What else are we to expect that we would conclude in our initial reading of that text? After all Genesis does say:

"and god said be light made and light was made and god saw the light that it was good and he divided the light from the darkness and he called the light <u>day</u> and the darkness <u>night</u> and there was evening and morning one <u>day</u>"

What else is a child to imagine the text means after reading terms such as "day" and "night" that we think we understand so very well?

What is Creation? And did it occur? Yes, obviously it occurred or it would not be here for us to study. In our modern imaginations we have all sorts of far-reaching thoughts and theories of alternate and parallel universes, and also various philosophies stating, or at least questioning, whether Creation is actually here or maybe that it is all only imagined by us. While we can play these sorts of mental games, in reality they do us little good. If you get seriously cut on something you will bleed. You can imagine that your life is only being imagined, but doing so won't stop the bleeding as you will promptly find out if you allow the bleeding to continue unchecked. Now of course bleeding from a cut does not "prove" Creation to be real, but bleeding does help to sway us from foolhardy destruction. Because we see it and stop it, we know it is real, and our alternative is death from bleeding.

So the real question isn't "did Creation occur?" Rather, the real question is "what actually occurred to cause Creation?" And that is where the debate sits and has sat for millennia.

Many people would be surprised to know that the big bang philosophy is nothing new, though the term "big bang" is relatively new. The debate of how Creation came to be has been raging on for thousands of years. The big difference between the thoughts recorded in history of those who lived long ago, versus our thoughts today is that we have actually been in space and can also observe things at a vast distance using very powerful telescopes, rather than only having our earthly naked eye perspective. This makes us able to better prove that, for instance, the Earth does in fact revolve around the Sun. Our mental box has been trying to be expanded for many years, but we humans always fall into the same old small confined religious and scientific mental boxes that we find so very comfortable.

The big bang has become one of the smallest and tightest mental boxes to ever exist, and it sits on the same shelf with the six-twenty-four-hour-day Creation box. Neither of these two boxes are useful to any of us.

An Interesting Point About Science, Scientists, and the Bible

It might appear as if I am picking on science's big bang theory, and I am, but I actually take greater issue with six-twenty-four-hour-day Creation theory. Both of these beliefs are absolutely absurd. Then to add folly upon folly, some believers, while not being able to mentally thwart the big bang theory, will accept the big bang as the hand of God Creating. By doing so, they imagine that they are cleverly blending science and religion, but they have not.

An interesting thing about science, scientists, and the Bible is that scientists typically refuse to look at the Bible through a true scientific lens. This is where big bangers will often point out that a Belgian Catholic Priest concocted the big bang theory. They do this in effort to diffuse Bible believers and stop them from pushing back against their big bang ideology. It is peculiarly evasive and ignorant to utterly reject the Bible that has so much back-up evidence, such as the places and people mentioned in it, as well as the fact that there are many other supporting documents that are non-Biblical. Additionally, the Bible is well received in many cultures, and is at least to some extent followed by most of western culture. To ignore this and discredit the Bible and then to go forth and spend billions to do something as insignificant as trying to prove that the Higgs-boson exists, commonly referred to as the "God particle", is quite ridiculous.

So how do we "scientifically" study the Bible? To start with, we use the same assumptions that we used to theoretically prove that the Higgs-boson exists. First, we believe that the theory is likely accurate, and we then proceed to prove such with any financial resources we can get. But this is where the problems with religion come in. Often, religion is much like modern science in that it refuses to properly scientifically examine *all* of the data given.

Take Genesis One for instance: We can wrongly assume that "earth" was made in six twenty-four-hour days, or we can see that it was there before any "days" were mentioned; but in this we still need to understand what "earth" is. And that particular point is critical in understanding Creation. If we were to apply the same level of effort in religion as was used to "prove" the existence of the Higgs-boson, it would produce a dramatically different perception of the Bible. But most "science-minded" people have prejudicially chosen to be ignorant of the actual words and the order of those words in the Genesis One Creation account. This is the exact same way that religious six-twenty-four-hour-day Creationists have prejudicially chosen to believe that their own misguided perception is accurate and true because *they* believe that "The Bible says so."

Consider the varying language used in post-Reformation Bible versions from recent centuries. To grasp this in a better light, we must understand that life is not as it is because the Bible says so, but rather that the Bible is as it is because the Truth and facts say so. This leads us to "Truth" and what it is. *Truth* is the ultimate quest of science, though many scientists do not understand that this is so, and some even have a difficult time explaining what *Truth* is.

For many modern-day scientists, it is their mission to, and they have been programmed to, illogically deny that the Bible is accurate and potentially true. This is especially heinous because most people have never read it cover to-cover. And most of those who believe they have read it did so as a youth, some using a children's Bible with never actually having read it cover-to-cover. Of course, that is not relevant in a discussion of Genesis One. But it does matter in trying to ascertain each our own personal honesty on the subject regarding our own agenda.

To go a bit further, many of those who claim that they have actually read the Bible cover-to-cover had decided beforehand that it was all nonsense. They had an underlying agenda of discrediting the Bible *before* even picking it up.

When people attempt to discredit the Bible they are actually ultimately discrediting themselves by revealing their own ignorance of the subject and actual history and migration of peoples mentioned in the Bible. There are many things you can say about the Bible, but to say that it is all just a bunch of "fairy tales" is about as ignorant as a human being can choose to be. There are far too many places mentioned in the Bible that we know existed because the remains of those cities still exist. Then someone can say "Well, sure it is easy to write about something after the fact and make up stories about it"; while this is true, it begs one important point of consideration: If the places were buried for thousands of years then how would someone have known that those places existed in order to write about them?

Further, since we only knew about the places because of the Bible, we searched for them and found them, thus illustrating that the information contained in the Bible was at least to some extent accurate.

Now, we can still argue that some ancient writer knew of the place and wrote fantastical stories about it, and then only afterward was the city destroyed and buried. This is explained further in the book *Understanding The Bible - The Bible How-To Manual AND The Things We Don't See*. This is a possibility, but then we have to make the assumption that it was a common occurrence where people wrote fantasy about the places and then the places eventually were overrun, destroyed, and buried only to have Bible believers blindly follow the text and search for the fantasy ruins mentioned in the text and then proceed to actually find those ruins as described in the Bible. The Bible generally has two accounts of destruction situations. The first account is a warning of what will occur, and the second is an account of what occurred or became of the place after the event they had been warned about occurred.

Consider the same mental thinking applied to the Higgs boson: Let's say that someone invented this boson in their imagination (and they did indeed invent it that way) and then wrote stories about it after. They made up stories about its purpose and function. All of this was put into a book or papers and then later, overzealous scientists decided to find it, so they then subsequently searched for the boson with the attributes that Higgs asserted in his text.

The difference with the "boson" versus ancient Biblical ruins is that the ruins are real and can be seen with the naked eye, where the Higgs boson is seen with multibillion-dollar equipment. And it is only "seen" and interpreted by the believing scientists who work with the equipment that was designed to prove such theories. The finding of the Higgs-boson is still a bit sketchy.

I am not saying the Higgs-boson is not true or accurate or that it doesn't exist. But imagine how differently the world would look at the Bible if that much money, effort, belief, and manpower was spent revealing the things written in the Bible's Creation account in Genesis One? Imagine...

But it's better that the rewards are generally not financial in researching the Bible, because that would draw, like a magnet, the same sort of greed that we see in pop-science today. Religion has enough prejudicial researching without the lure of greed for money entangled in the research aspects.

Our problem is not trying the prove the Bible to be accurate, rather the problem is our misunderstanding of the Biblical and scientific data as we read and interpret it. The reason so many people have decided to abandon the Bible is that the book itself is often the basis people use to prove everything, rather than being the obvious historical confirmation of the proof that it actually is. If you view Genesis One as a purely scientific account of Creation, then you don't have to cheat and stretch the facts, you need only view it from a Creator perspective. We need to see things through the eyes of others and seriously consider other perspectives if we do not want to discredit ourselves in the long run.

Multiply by Dividing

Is there really a "god particle"? I suppose that would depend upon what "god particle" is supposed to mean. We each have to try to grasp what the Creator actually Created. As humans who are only accustomed to an environment with physical attributes, we have a difficult time trying to imagine utter and absolute nothingness. Utter and absolute nothingness includes no existence of the sub components of the three parts of atoms—the protons, neutrons, and electrons. This means there would have been no elementary particles or components. There would have been no baryons, bosons, gluons, gravitons, mesons, photons, W-particles, Z-particles or any other particles we can find or theorize. All of those would have been Created. There would have been no theoretical particles, no light, and possibly not even space itself.

In reading the first sentence of Genesis One, we have little to guide us in the potential meaning of that first sentence. From our life experience, the first reading anyone does of the first sentence of Genesis leaves us with little choice but to make the assumption that the "heaven" and "earth" mentioned in it are the *Heaven* being the place where Angels hang out or possibly space and "earth" being the planet *Earth* we all walk upon. However, when you read the rest of the Creation text then you have to rethink such simplistic childlike thoughts, and you are forced to look a bit more logically at the text that follows the first sentence. You have to consider the specific *order* of events and *at what time* something was named and *why* it may have been named as it was.

As mentioned in previous chapters, it is highly unlikely that "heaven" spoken of in the first sentence of Genesis One is "Heaven" with Angels or that it is space as we know it today, and it is also highly unlikely that "earth" in the first sentence is our planet Earth.

Scientifically, it is almost certain that "heaven" was first seed of preexistence of an abyss of eternal, and "earth" was first seed of lowest level of physical matter—or that which makes up those elementary pieces so small that we, up to this point in time, can only theorize about their existence as we set out to attempt to prove that existence. However, our ability to guess at what could have been the initial particles at Creation does not have any bearing on whether or not those particles where Created by a discerning being.

We often have some sort of warped idea that just because we can explain things it means that it wasn't intelligibly Created. Doing this is like a mechanic looking at an automobile and understanding it, but then disregarding the engineers who designed it and assuming that those engineers and builders never existed just because the mechanic now understands how it works. Yet that mechanic offers no viable alternate theory for the creation of the automobile.

We must understand that our opinions have no bearing on reality. What is—*IS*, and it is our mission to try to understand that "*IS*". The "heaven and earth" referred to in Genesis One logically had to be the most base building blocks of anything physical or tangible. They would have had no gravity, no mass, no weight, no pull, no push, no force of any kind. Based upon our knowledge of known particles, it would have been invisible and possibly some sort of fundamental energy that would likely not have been detectable with our current technology. You might question this as a bit far-reaching and it may be so; however, the text reads:

"in the beginning god created heaven and earth and the earth was void and empty and darkness was upon the face of the deep and the spirit of god moved over the waters"

The first two verses indicate that "earth was void and empty and darkness was on the face of the deep". We picture this in our planet-Earth-view as a barren planet possibly as a slurry of chemicals that are very deep, much the way we speculate some of our distant gaseous planets are composed as we see them today. Is this possible? Not likely, because at this point gravity would not have yet existed to gather clusters of matter together.

"and the spirit of god moved over the waters"

What are the "waters"? "Water" in ancient language can mean *many* or even *movement*. There are many substances that people work with in industrial environments that when you handle them, these substances are handled like water because the particles are small and are able to be poured, yet they are not wet like water. If you inject a bit of air into most powders they behave much like water causing them to flow and making them able to be poured much like water.

The point is that the term "water" having been translated, likely originally meant fluidity or ability to move or flow. There would really be no true comparison, but the idea of a fog might suffice for some people, though it would likely not have been a visible "fog". This is to say "heaven and earth" of Genesis One was made with fluidity, or we can think of it in terms of a transparent fog of pre-subatomic matter, and it may have been an endless single item in the location of *everywhere*.

Then God said "'be light made' and light was made". This event would have been the first division of this invisible fog of pre-subatomic earth-matter waters. Through this "heaven and earth" matter being called upon and the fog broken into energy or particles would have been multiplied to a value that we have a difficult time comprehending. Yet it may have only been a slight variation in state. Scientifically, we have no reference point for such an event, but we are slowly discovering ever smaller bits of matter.

Douay English Genesis 1:4 "And God saw the light that it was good; and he <u>divided</u> the <u>light</u> from the <u>darkness</u>"

Douay Latin Genesis 1:4 "Et vidit Deus lucem quod esset bona : et <u>divisit</u> <u>lucem</u> a <u>tenebris</u>"

Scope of View

Our scope of mental reference, such as "universe" and "galaxy" and "solar system" can be somewhat ambiguous depending upon who is considering those terms. "Solar system" seems reasonably understood to people in the science fields, but other people might not grasp that every star is itself a solar system, likely with items orbiting it. The term "galaxy" is also reasonably clear in the mind of most people involved in science and astrophysics, yet many other people will confuse the terms *galaxy* and *universe*—a galaxy being a massive cluster of stars or suns along with their orbiting planetary matter.

"Universe", on the other hand, is the one term that even some science people have a difficult time grasping. For instance, when we speak of "multiple universes" we defeat the idea of the term uni-verse. Think of "uni" as one, uno, or like uni-cycle a one wheel bike. "Multiple universes" is an oxymoron. Universe is meant to be an all-encompassing term that includes *all* of Creation. We must resist having our limited minds keep building such universe walls only to invent another new wall that we need to break through each time we knock down the previous wall. The term "universe" should have no walls to break through because it has no known or visible end.

Our mental scope of view interferes with reality and that is our most prominent shortcoming as humans. We are letting our preconceptions block our progress of understanding.

Chapter 8

Is the Church a Building?

When we look at Genesis from a *scientific* viewpoint we are forced to see things we would otherwise not see. Yet we struggle in doing so because of our prejudicial reading of the text.

Consider the "Church". What is it? It is a building? Is it the Catholics, or some other religion? When you read the Bible, you should eventually realize that the "Church" is the people (See the book *Understanding The Church - Upon This Rock I Will Build My Church*). But in our culture with so many people falling away due to misunderstanding the Bible, the "Church" has become buildings plus the Church leaders who occupy those buildings. Then when a few unsavory people who happen to be occupants of some of those church buildings behave badly towards their fellow man, an entire people gets discredited through that bad behavior, thus turning evermore people away from the Church and away from the Bible.

Let us be clear that the Bible has nothing to do with what we today call "religion" or the "church". Rather, understand that religion today has become what it is, be it misguided or of perfect harmony, because of the Bible and the past, not the other way around. The books of the Bible came first and are not responsible for *our* cruel actions, ignorant interpretation of events, beliefs expressed, or attitudes about the texts.

When one religion doesn't work for a person we have a tendency to find a new religion that agrees with what we want to believe, which is often driven by a specific priest or preacher. Consider what happened with the Reformation and all of the religions that have sprung up from it.

The point is that we gravitate towards what we want to believe scientifically and religiously. We, as a people, are somewhat formless and void when we choose to ignore obvious facts. If we ever hope to understand how the heavens came to be we must move our minds out of our current tiny constricting mental scientific and religious boxes, and guide them into the vast open Universe. There is nothing wrong with religion or science when we allow ourselves to see in an open and unbiased manner.

Void and Empty or Formless and Void

Genesis One's second verse says "and the earth was void and empty" or some versions mention "formless" or "without form". This indicates that the "earth" was not a ball of dirt such as we walk upon today, because "earth" was without form. "void and empty" would also indicate what was spoken of in a previous chapter regarding a transparent/invisible base-matter or energy. It is something, and it is there, but it had not yet been substantiated. You have likely noticed a Latin version of the parallel text that has been included for your convenience. This is included simply to demonstrate that some of our translated words have phonetic similarities and for other words they have used complete replacements.

And then:

Douay English Genesis: "god said let there be a firmament made amidst the waters and let it divide the waters from the waters and god made a firmament and divided the waters that were under the firmament from those that were above the firmament and it was so and god called the firmament heaven and the evening and morning were the second day "

Douay Latin Genesis: "Dixit quoque Deus Fiat firmamentum in medio aquarum et dividat aquas ab aquis Et fecit Deus firmamentum divisitque aquas quae erant sub firmamento, ab his quae erant super firmamentum Et factum est ita Vocavitque Deus firmamentum, Caelum et factum est vespere et mane, dies secundus"

By now you probably have a good idea of the formless and void part of what was being referred to as "earth" in verses one and two. Moving along to the divisions in the text that we began getting into in the last chapter:

"and he <u>divided</u> the light from the darkness and he called the light day and the darkness night and there was evening and morning one day and god said let there be a firmament made amidst the waters and let it <u>divide</u> the waters from the waters and god made a firmament and <u>divided</u> the waters that were under the firmament from those that were above the firmament and it was so and god called the firmament heaven and the evening and morning were the second day god also said let the waters that are under the heaven be <u>gathered</u> together into one place and let the dry land appear and it was so done and god called the dry land earth and the <u>gathering</u> together of the waters he called seas and god saw that it was good"

Notice the *divisions* and the *gatherings* stated in the text. When we stick to our narrow Earthly view of this text, it is confusing and arbitrary. But when we step outside of our preconceived Earth-view boxes and look at it with open minds from a truly scientific perspective, it gets very interesting scientifically and religiously.

When we view "heaven and earth" in the first sentence of Genesis as a sort of pre-subatomic matter then these divisions and gatherings would indicate an attraction type of activity as used in the term "gather"—or what today we call gravitational forces. Some modernized paraphrased versions of the Bible have the *divided waters* translated as "clouds in the sky" and the *waters under* being the "oceans". But that is a bit of a stretch when reading the text in an honest and accurate manner. Such narrow perspectives only serve to nail our mental boxes shut ever more tightly.

Formation of Heavenly Bodies

In the following text there was a "firmament" made and it was named "Heaven". Now this particular "Heaven" is possibly the Heaven with Angels in it, but also, at the same time, it can be our Outer Space as we know it today, being empty or void containing the emptiness of only a pre-subatomic matter substance or form of energy. This pre-subatomic matter substance is something that we would likely not be able to detect with any of our modern instruments or methods at the point of Creation.

"let there be a firmament made amidst the waters and let it <u>divide</u> the waters from the waters and god made a firmament and <u>divided</u> the waters that were under the firmament from those that were above the firmament and it was so and god called the firmament heaven and the evening and morning were the second day"

The waters that were above were divided from the waters that were under. This is a good description of the effects of what we call "gravity". Then:

"god also said let the waters that are under the heaven be <u>gathered</u> together into one place and let the dry land appear and it was so done and god called the dry land earth and the <u>gathering</u> together of the waters he called seas and god saw that it was good"

The text just shown is a good indication that this was a likely result of a force, similar to our modern day understanding of gravity.

But now we have an additional reference point to consider. Is the text where "Earth" and "Seas" were named referring to *our* tiny little planet Earth and *our* oceans? Or is something much more encompassing occurring here?

The likelihood that this is referring to all of the heavenly bodies is the more likely reference point to view this from, with the idea that the named "Earth" now being a substance we here on our planet recognize as dirt or elements. Let's consider the images we have that were taken by the satellites orbiting Mars. Whether it was water (H2O) that caused some of what we see or some other wetting liquid, we still recognize this as very similar to types of land formations that we encounter here on our planet Earth. The point being that the particular reference of dry land being named "Earth" and the gathering of the waters being named "Seas" is likely true throughout all of Creation, that is to say the entirety of the Universe.

A pre-matter substance was somehow catalyzed to cause grouping or attraction of some sort. This part of the subject is somewhat ambiguous because there is very little detail given about it. Yet it is interesting that the two things that are most evasive to our understanding, that is to say the *Creation of light* and likely the *Creation of what we call "gravity"* are so close in nature—as in $E=mc^2$—just happen to be the first two Creation events from day one and day two. They are essentially the beginning of what we would scientifically think of when considering anything at all actually existing in any sort of scientifically detectable way.

Day one was the Creation of light and day two likely the Creation of sub-atomic particles or elements, thus causing gravity to be able to occur and things to actually exist that are scientifically detectable by us today. Then on day three the waters (not H2O) were separated to above and under states, or superior and submissive.

After that point the "waters under were gathered into one place making dry land appear" and then they were named. So as human beings who live on a planet we call "Earth", the question we must ask ourselves is, is this particular event exclusive to our planet or did this more likely occur throughout all of Creation, thereby forming planets and stars throughout the entire Universe? If you study the Genesis One text from this perspective it is very logical to suggest that the day three events caused the formation of *all* Heavenly bodies to *begin* to form after that sequenced event.

Insertion Theory

Regarding the formation of heavenly bodies or the large clusters of heavenly bodies that we refer to as galaxies, there are varying theories as to whether this changes much or if it is in a near constant or static state of being. "Static state" theorizes that not much changes, and the Universe neither expands nor contracts.

With regard to the big bang, the thought is that it is ever-changing and expanding. But alternate big bang theories project a maximum expansion at which point it would eventually begin to collapse again into a subsequent singularity. Then continuing to "big bang" once again to a maximum expansion where it would eventually collapse again, thus oscillating from one state to the other, endlessly, over vast amounts of time.

Regarding expansion of the Universe, "Insertion theory" is where new galaxies are created in between the expanding distances. Is this possible? That's a tough question to answer without any solid information, but if Creation is essentially completed, then likely not. However, it is possible that formation could occur given that the elements tend to be drawn together due to their gravity. In this case we have to try, in our human minds, to agree on what a galaxy is, or for that matter, what a star or a planet is. Could galaxies or stars still be forming all across the Universe as this book was being written?

Here we get into a problem with the text because it indicates that the Creator rested on the Seventh day indicating that no more was, done. But is this really what the text is indicating? Not exactly.

I am not saying the Creator continued creating, but rather that the Creations continued. If you start a chain reaction by adding chemicals together, then your part is done once the mix has been made, yet the reaction keeps on working. You just sit back and watch. The same would be true of the Creator's work. The

Creation of light and other elements cause gravity and then gravity does the work autonomously, slowly creating all of the heavenly bodies over long periods of time. Given the vastness of the Universe, this gravitation action could be constantly occurring forever as it slowly gathers the "waters", eventually causing dry land to appear such as planets that orbit their star.

Many Mansions

Just as there are supposedly no two snowflakes exactly alike and no two fingerprints exactly alike and no two leaves exactly alike, there is also likely no two stars or planets exactly alike. Of course we don't know this for sure, but while things can be very similar, they are typically not perfectly identical.

At this point you should be wondering: If this is true, then wouldn't that mean that there are other Earth-like planets out in space, planets that are similar like snowflakes, yet different? The answer to that is probably, yes, without question. I have a difficult time imagining that all that we see in space is dead planets like Mars or our Moon. There are likely many mansions or jewels like our planet "Earth" made of the same substance referred to as "earth" in Genesis One where dry land has appeared that are orbiting most stars in space.

Chapter 9

Protective Nature of Creation

Creation is an interesting topic that is both narrow and vast. It is narrow in that Biblically speaking we have little information to go on. We must check to see if the little information that the Bible offers fits with our true scientific findings that are actually found and are more than just speculations by overzealous researchers and theorists. But on the other hand, the Creation topic is also vast because there is so much more to it all.

Can you imagine if the Genesis Creation account actually went into the deeper aspects of the elements? Try to fathom attempting to explain the periodic table to anyone who has never been able to hear of this sort of scientific dialogue. Initially, it would be like explaining it to a young child. Some of the less-discerning people from thousands of years ago might believe it, but they would likely not really understand it. And what sort of spirits or demons might they attribute these things to that are obviously not demons? So, we have to realize that since we, today, struggle to grasp the brief Creation account that is offered in Genesis, we would struggle even more if not for our relatively recent collective encounters with science. We also have to consider that the Genesis One text would be very, very long. Most of us are too lazy to read the Bible as it is, now imagine Genesis One's first six days explanation alone being as long as, or more likely much longer than, the entire Bible.

We are given only what we are able to handle when considering our human nature, and we will likely not get more until we learn to handle and understand what little information we currently have, which is a lot more than some of us actually want to handle. In about six thousand years of clearly recorded human history, we have failed to even get the first page of the Bible correct. If we can't figure that out, then do we really deserve to advance? Think about that.

The limited amount of information we receive is a protection for those who receive the information. There are certain things you should not tell little children because they will not understand the dangers involved, and out of their wonderful curiosity, they will often try the things that they hear or see, sometimes causing themselves great harm or toil. Countless lives have likely been saved due to the limits regarding the little documentation we have been given in the Bible about Creation. Consider the discovery of the powers held in an atom and what we have been able to do with it. Sure, it's neat to release that much power so suddenly, but consider the utter danger from an atom bomb. What evil would we humans unleash if we knew even more secrets about Creation?

We do not deserve to know some things until our world-culture is wise enough to stop killing each other. And for the immediate moment, we have been failing miserably at that particular task.

Protective Forces of Nature

Just as we are protected from ourselves by the limited information we were given, so too are we protected from injury by the limited radiation we receive due to Earth's atmosphere.

Nature is not very well-defined in our minds because we have natural things like trees, and we also have the nature of our minds, and then there is human nature. In our international world, for several decades we been have attempting to be more nature-oriented, as in trees and grass, in effort preserve our planet, yet in doing so we are always trying to stop "nature" from doing its work by us interfering with its "natural" changes. For instance, when the Great Lakes hit a fifty year high-water mark, people often place rocks at the lake's shore to stop erosion. There is nothing wrong with protecting the shore, but doing so is certainly interfering with the natural erosion that would otherwise have occurred.

The point is that nature will do its thing whether we are here or not, and we are the beneficiaries of that nature. For a very long time our atmosphere has been doing a wonderful job at allowing us to breathe, which is made evident by the fact that we are here and we breathe. We also have the planets around us, and those planets and their moons and our moon are attracting asteroids like iron nails drawn to a strong magnet, asteroids that might otherwise impact our planet Earth and snuff out life in large areas of our planet.

One question we humans seek to answer is: Is this all guided by the Creator or is it just random occurrences of nature? If you look at the systems that are currently in place, it is apparent that Creation was put in motion with many safeguards and life was initiated on habitable heavenly bodies. The other bodies in the same celestial neighborhood then help to keep balance and work to clean the surrounding space. It is truly a beautiful system when you really think about it.

Did God Create the Earth to Look Old?

An earlier chapter mentions that some of those who theorize that creation occurred in six twenty-four-hour days have in the past claimed that God made the fossils as a part of that initial creation. In other words, God made the Earth to look old in their view. It is highly unlikely that a Creator who has the ingenuity to invent *matter* would waste any time creating false fossils in an effort to "make things look old" so that we could mentally reconcile geology with the Bible. When we use such absurd rationale, we dupe ourselves and keep pounding more nails into the lids of our already tightly sealed limited, tiny mental boxes.

I am not going to go any further down the fossils path because it touches on the *Creation* versus *evolution* topic and it is not what this book is about. There is much more detail about that topic in *The Science of God Volume 5 - Boats, Floods, and Noah - The Deluge*.

Did God need to create the Earth to look old? No, the Creator did not need to make the Earth look old, because the Earth is old. But while evolutionists who read this might be cheering at that statement, much of the fossil discoveries found are not accurately analyzed by most scientists of the past few centuries. If the Bible is at all accurate, then modern scientists might have some explaining to do. The question is not if the fossils are old, the only real question is how old are they? The Earth is very old as is indirectly stated in the Bible, it's just that there are too many people who refuse to see it because of their preconceptions due to misreading the text in the Bible, most notably in Genesis One verse one.

The Earth is indeed very old and the fossils tell a great deal about our planet Earth's history. The question is: Are we reading the fossils accurately? And just how old are they? But that is for the volume of *The Science of God Volume 5 - Boats, Floods, and Noah - The Deluge*.

Principles

In the Bible Jesus talks about not pointing out other people's faults, as he was telling us to "remove the plank from our own eye before trying to remove a small sliver from our brother's eye." In other words, we all have our own biases and errors and therefore before we condemn another we should make sure that we have removed our own errors from our lives. I would assume that most people can agree to this—Most people, but not necessarily all people.

If we are to ever find the truths of Creation, we have to stop unjustified and vicious judgements upon other people and their theories. But we also have to take a stand when the theories just don't add up. It is irrelevant regarding whether their theories are of good intentions or not, because when we are wrong then we are **wrong**. And if we insist that we are correct and we lure others into believing our erred theories are true, then we become false preachers who are leading people astray. In the case of the Creation account, we can actually lead them away from the Bible and away from the God spoken of in the Bible when we inaccurately interpret the Genesis text and/or if we inaccurately interpret science.

Here is the part that is perhaps the most injurious to people: "believers" who are erred in their theories are likely responsible for turning away more people from God than is any atheist-scientist or college professor with their promoting of their highly doubtful theories of big bang and human evolution. The problem with outspoken atheist scientists and outspoken college professors is that they impede further research down alternate lines of thinking that might otherwise be done by their students, thus hindering *true* scientific study. But if the students where not led astray in their beliefs about Creation by their preachers and priest and parents, then they would have the tools to present sound debate when in discussions with outspoken atheist-scientists and/or outspoken atheist college professors.

Our preconceived comfort-points satisfy our desire to understand. Our comfort with our beliefs outweighs our ability to bear the burden of what we don't yet understand or don't want to have to try to understand. We prejudiciously deny the information in the Bible and fail to test it and then we go onto believe inaccurate information proposed in aspects of the big bang and evolution theories, or in the six twenty-four-hour-day creation theory.

Another principle point to keep in mind is that we must keep a division between the topic of evolution and big bang. Christians or other believers have a tendency to blend these two aspects of Creation. The two aspects are not big bang and evolution, but rather are astrophysics and microbiology. While both occurred during the Creation accounts in Genesis, they are very different discussions that connect only at the point where micro biology began at which point the astrophysical aspects were already largely in place. (See *The Science Of God Volume 3 - Day Five and Day Six - The Creatures - Revolution or Evolution, The Science Of God Volume 4 - Day Six - Evolution versus Man - In Our Image, The Science Of God Volume 5 - Boats, Floods, and Noah - The Deluge.*)

We Can Rationalize Anything

Through our preconceived thoughts and our desire to understand, and our desire to be right, we can, and often do, rationalize *anything* to suit our lust to not be proven wrong. Don't let yourself get caught in that trap. We must consider all alternate theories and disregard those that cannot stand up to any scrutiny.

Our Passionate Arguments

Regarding our unsubstantiated beliefs, when we have a deeply rooted belief we tend to stand our ground and will often fight to protect that belief at all cost. And, as history has repeatedly proven, we will do this even unto the point of death.

But if our belief is incorrect, *then it is **incorrect**.* Wrong is wrong and it does not matter how much you believe something is true, or how loud you shout, or how many names you can call your opponents in the discussion. In the end, if you are wrong *then you are **wrong**.* Often, the so-called "science community" scoffs at the belief of the religions with regard to Creation, and often for good reason. But what they fail to see is that their own beliefs are no more accurate and are no less "beliefs" than what many creationists often propose.

It is of no consequence to Truth who you are, or who you think you are, or if you have your religion, or if you insist on an inaccurate analysis of the Biblical data—Wrong is **wrong**. If this is your approach to "science" then please work to silence yourself until you have better understanding. Do not try to convince the big bang scientists of your erred beliefs about the Bible. And do not attempt to convince Bible believers of your erred big bang theories. Doing so will only make yourself and science and the Bible look bad in the eyes of others. It also makes people who understand science and people who understand the accuracy of the Bible appear foolish in the eyes of onlookers, because, sadly, loudmouthed-error tends to get far more attention than it ought to.

The Bible is often misinterpreted by people who think only *they* are able interpret it as they proceed to spew their uninspired erred preconceptions to the world. You might imagine I am speaking of some erred preachers of the Bible, but it is far more than that, the same is true for many in the science world. The Bible is interpreted in their eyes as if it is in error because of what *they choose* to believe it says or means.

It is what we do with the information and how we analyze it that will protect us from our own folly. If we fail to filter the information through the process of truth, then it will likely be our downfall and we will succumb to error and lies. It is the skill of filtering through truth that offers you protection of self and advances true science, and will allow us to begin to actually understand Creation.

Chapter 10

Universal Expansion

We lightly touched on universe-expansion in an earlier chapter. Here we are going to get into possible types of expansion that have been theorized to have occurred during the initial events of creation whether godless or not.

Things can expand in several ways. One widely discussed expansion type is that of an explosion like is asserted in the big bang theory. In this type of expansion everything is moving outward from a central point. In this theory there are sub-theories regarding the speed or velocity of the expansion and whether it is constant regarding it speeding up, or possibly even slowing down.

Another common expansion illustration is in terms of a balloon expanding where everything gets further apart on the surface as the outer skin of the balloon expands. It is not a very good example, but it is something that you can easily test on your own by making marks on an empty balloon and then blow air into it and watch as distances between the marks increase.

An even simpler illustration of the big bang expansion would be to take a long rubber band and cut it so that it is no longer a loop and then mark it at even intervals. Then as you stretch it, you can see the distance between the marks increase, or decrease as you release the tension. An explosion big bang and a balloon and a rubber band are all inflation type expansion.

Another type of expansion is growth expansion like we see in nature every day, such as a tree growing larger and larger every year, or a plant such as a flower growing. A tree is a good example of this sort of expansion because it keeps repeating a general but random pattern of branches and leaves, similar to the way we see general patterns in the heavens with stars and galaxies that all appear different, yet we can clearly classify them as stars and/or galaxies.

There are a couple of questions with regard to Creation and expansion, and they are as follows: Is Creation expanding still now to this day? Or we can ask, has it even ever expanded in any way at all?

The first question assumes that it has expanded which is more commonly discussed in science than in religion. However, if you consider the second question, then the first question could be considered a leading-question. We will likely never be able to answer the expansion questions with any certainty because, as far as we can see with our telescopes, there seems to be no end to space and no end to the universe within space. Logically, we can assume that if we could travel to the most distant galaxy that we are currently able to see, then we would likely see more of the same. Is it all infinite?

Big bang basically places us at the center of the explosion, or else we would see a very different view depending upon to which direction we looked. The book *Bending The Ruler - Time Travel, The speed of Light, Gravity, and The Big Bang* goes into more detail about expansion anomalies in big bang theory.

While a rubber band and a balloon are good fundamental illustrations, they both place an imaginary person on a plane and when that plane stretches, everything moves somewhat equally away from them. However, in a 3-D model you can place a person at the center of a spherical explosion or expansion causing a very different perspective. As is proposed in big bang theory, we can theorize that any selected smaller area between the furthest edge and the central point of ignition is stretching. In reality, an actual explosion would produce a very different result than what is proposed in big bang, but for discussion sake we will continue the thought.

Just as a rubber band will increase the distance between any two marks equally and double that distance between any two marks when skipping a mark in between, so too will most of the surface of a balloon. That is *most* of the surface. But in such inflationary theories it is better to examine the drip end of the balloon that is typically opposite the air inlet. In this area of the balloon it is more like the expansion pictured in big bang. When the balloon is completely empty and many dots at equal distance are placed on the balloon and the balloon is then inflated, you will see a very different expansion result as you near the drip area. The drip area is the area that the balloon material dripped off during manufacture where it gradually gets thicker. If the balloon is inflated, you will notice that as you look nearer to the drip spot that the dots will be closer together. A balloon's drip-off spot is a more fair depiction of big bang than is the marked rubber band example.

Regarding the rubber band example, we would be extrapolating that example throughout the entire Universe, but evenly in all directions. An actual bang-type expansion would leave a void in the center area much the way a real explosion forces everything away from the ignition point, or the distance between each marked interval on the rubber band would increase evenly even as the distance from explosion center increased. Some people use an example of baked goods expanding, such as bread dough with raisins in it. In the raisin bread example, the illusion is given that the raisins caught in the dough all separate from one another at a similar rate and that this is a good example of how that view of expansion might occur. But this poor example overlooks the point of fact that the raisins still all separate in a radial manner from a center point of equal resistance. Bread rising is in essence a slow-motion explosion expanding in all directions, but since it is actually not violently exploding like a true and typical explosion, it is a very poor illustration of the reality of physics in that regard. An explosion will leave a void in the center, especially if it is still expanding. Logically and scientifically, it cannot be both ways. We see the explosion-void-effect in space when looking at some of the more interesting images where a ring of debris and/or radiation forms around where the initial explosion likely occurred. But while a particular galaxy might show explosive expansion, the Universe does not show explosive expansion as far as we have observed at this point.

Tree-like growth expansion is far more likely to have occurred than was any explosive expansion in the initial Creation events prior to the stars and planets being formed. However, as difficult as it is for us to accept, there may have been no expansion at all—*ever*.

Our human experience is only familiar with a few different expansion types such as explosive expansion and growth expansion. But it is possible that the initial sub-matter referred to as "earth" in the first sentence of Genesis One was everywhere simultaneously. And further, it is possible that the formation of the heavenly bodies could have only occurred after light was made to be and while the "firmament" was also occurring simultaneously throughout all of space. This does not mean that it all occurred in an instant or in perfect harmonic timing, but rather that this all only *began* when gravitational events began doing what gravity does from the very smallest to the largest.

Some heavenly bodies would have likely taken longer because the random pull of the surrounding matter could have broken anywhere and attracted to one body or another in a random manner. No true time-index was given for this series of events in Genesis, but we can assume that since God is believed to be everlasting and to have always been there that there is no regard to time in what we think of as days and years.

In reference to time and "years" that we humans are familiar with, there could have been thousands, millions, billions, even trillions or more of those increments during Creation. And from our human timing experience, the counting of time as we know it could not have occurred until at least the fourth day when:

"god said let there be lights made in the firmament of heaven to divide the day and the night and <u>let them be for signs and for seasons and for days and years</u> to shine in the firmament of heaven and to give light upon the earth"

Is Heaven Real?

We are going to briefly step outside of astrophysics and touch on metaphysics for a moment. Is "Heaven" real? In this case we are not referring to the "firmament" that was named "Heaven" but rather the Heaven of angels and souls. Is this Heaven of souls a Created place? Is it a state of being? What is it? One of the things that we fail to see as humans is that Creation could have more than one "level" for lack of a better word. In other words, we might be in a world that is tangible to us as humans but there may very well be another somewhat parallel existence that we cannot detect because it could be entirely different. Could it be that "god made a firmament and divided the waters that were under the firmament from those that were above the firmament and it was so and god called the firmament heaven" where the waters above the firmament are what we consider to be "Heaven". And "above", or "Heaven", is where souls reside? It is something to think about.

This brings us to the question of the directional indicators "above" and "under". Are these *directions* as *we* understand direction to be today, or could "above" and "under" be intended to mean something else such as *less than* and *greater than* or *less powerful than* or *more powerful than* the firmament? Is it possible that the waters under the firmament are subordinate to the firmament and the firmament is able to affect them in a gravitational manner as is described on the third day?

"god made a firmament and divided the waters that were under the firmament from those that were above the firmament"

Waters that were under were somehow separated from the waters above, and then the waters above are no longer referenced in the text of the first four days account. In a scientific analysis of the text, the directionals *under* and *above* are irrational unless they are **not** directionals as we assume them to be in our typical human perspective of directional understanding, but rather are more closely associated with their susceptibility or non-susceptibility to the "firmament" or "Heaven".

We tend to want to picture a physical gap or distance between these "waters", but it might be more of a coagulation of the lower waters. Think of the waters (waters, plural) in terms of a liquid solution where it is liquid much like water and then a firmament is added or a type of catalyst that causes parts of the solution to coagulate. Yet they are still both in the vessel of space but some of the waters are coagulated and are floating within the other waters. They are separated from each other by the firmament that was added. Compare it to water with metal filings in it, where if you place a magnet near it then the gravity of the magnet affects the iron filings but not the water. The iron filings in that case would be the "waters below".

Now I am not suggesting that a liquid-like solution was there at Creation, but rather I am using this as a rudimentary example to show what "under" and "above" could possibly and likely have meant on the second day, since any sort of logistical direction at that particular time would have had no relevance until that particular point in Creation had started or was completed. What actually became of the waters above? Scientifically, the waters cannot be the clouds or the rain because those are from, and thus are a part of, the Seas that are the result of the lower waters being gathered when dry land appeared.

Is there another aspect of Creation involving the waters above that we cannot detect? Could the waters above have become what we refer to as "Heaven" with angels and all?

Some scientists theorize parallel universes that reside in the same space as we reside, yet have a difficult time accepting the concept of Heaven where our souls go after death. It seems to me that, scientifically, we all need to revisit Genesis One with an entirely new perspective.

Calculating Time Using Stars

The theory of an expanding universe is a bit of a stretch if you'll pardon the pun. Using stars for index position reference purposes and calculating time for long time periods assumes that all things are constant while at the same time claiming a moving universe.

There are too many outside factors to make absolute assertions regarding the "age" of the Universe. The Universe is likely very old in human years, possibly billions of years and maybe even trillions of years or more. "Redshift" dating methods can likely not be relied upon. Any theory that is built upon an expanding universe, especially an explosive bang type expansion, will carry in it the same anomalies that the theory that it is built upon carries.

Here are things to consider when trying to calculate time through redshift: We are not certain of how the gravitational effects of heavenly bodies, or groupings of them, will affect the light coming to our eyes or telescopes. If the light traveling from a distant galaxy for billions of years must pass anywhere near the gravitational pull of a galaxy that lies between the distant galaxy and us, then that galaxy could be quite far out of alignment between us and the distant galaxy that we are viewing, causing a sort of prism effect resulting in redshift.

Light obviously behaves differently based upon its wavelength as shown when sending it through a prism, so we can assume that the same can be true in space, especially at vast distances. The likelihood is very high that a massive but gentle pull from in-between galaxies or stars affects the light of the galaxy being viewed by us by separating the light somewhat like a prism does.

We also have to consider that just as our planet Earth revolves around the Sun, so too does the Sun likely revolve around the center of our Universe, and our ability to detect this movement is negligible. Now consider a distant galaxy that is lightyears across and us viewing it. The movements or objects within that galaxy might be moving hundreds of thousands of miles per hour. But even at the speed of light of 186,000 miles *per second,* it would be hard to detect the movement. If some galaxies are as large as we believe them to be, and are tens or even hundreds of thousands of *light years* across, then a star in that galaxy moving at the speed of light for an entire year would move only one one-hundred-thousandth of the way across the galaxy in an Earth year.

Now when you consider that the images returned from our orbiting telescopes might only have a handful of thousands of pixels across, each pixel would then represent maybe ten to twenty light years distance. So, if we could watch that same galaxy at full frame resolution on the camera it would take ten to twenty years for the star moving at the speed of light to jump to the next pixel in the image. There is a bit more to it than that, but this basic example should give you an idea of the scale and distances we are attempting to speculate on in our theories.

There are a couple of reasons I mentioned the point about speed and distance within a galaxy. The first is that some people wrongly believe that the heavenly bodies are static and that space remains the same—always. But with our modern telescopes and just sheer human logic we can tell that this is not so. Everything from the smallest particle to the largest body in our solar system is in motion, and the same is almost certainly true of every other star and every other galaxy that exists. The second point is that movement as stated in the big bang theory would have everything moving apart relative to any heavenly body grouping. Since even movement of distant bodies at the speed of light would be difficult to visually detect, we believe we are only left with redshift methods with which to estimate speed and distance. When viewing for "redshift" some stellar groups show progression and some show "regression", which is somewhat contradictory in a theoretically expanding universe.

On a more local view, within our own galaxy, we cannot reliably use other stars for very long-age position because their procession or movement could have been altered due to a super-massive asteroid collision causing a change in current position of the star. And further, gravitational effects of nearby stars are also altering the natural course of any specific star. This would alter its position from where it would have been if left undisturbed and thus would change our estimations of time for certain calculations, for instance, what the night sky specifically looked like many thousands of years ago.

Mass over Velocity

When something is traveling in space, the item's speed is both relevant and irrelevant. Speed, or velocity as it is referred to in science, in a sense does not really exist in space for any single body. However, that velocity becomes an important factor when there is more than one item in space. If space was all empty and there was only one item in the entirety of space, then there really is no such thing as speed or velocity. This is because you would not be able to tell if the item is moving.

However, once a second item is introduced into this mental exercise, it forces velocity into the picture the moment the two items have any change in direction relative to one another.

The same is true regarding mass. You can only begin to detect mass if more than one item exists.

Both mass and velocity are references of the difference between two things.

It is incorrectly theorized that the energy and mass of a particle will increase with its velocity, so a grain of sand impacting another object at the differential speed of light could potentially make a very big explosion. However, this has a few additional real-world circumstances to consider. Different materials will behave in different ways, so the specific composition of material that the grain of sand is made of, and the type of material that the grain of sand is impacting, will greatly affect exactly what occurs upon impact. We also have to consider form dynamics. Consider a projectile made of lead such as a lead bullet with a point hitting water at a high speed, then imagine what might happen if that lead bullet is first made flat like a coin. If the coin shape hits the water on its face, then it will behave much differently than when it has its point and is shaped like a projectile.

Movement in space causing any large collision in space is going to seem violent to us. We are amazed at the speeds we reach on here on Earth or when we send rockets into space. Now, imagine something moving ten or more times the speed of a rocket. Or try imagining the speed of light. A rocket moves at roughly 20,000 miles *per hour*, but light travels at approximately 186,000 miles *per **second***. That means that light is traveling roughly 33,000 times faster than a rocket.

Murmuration

We see murmuration when a large flock of birds is flying together and they suddenly change direction. The birds don't collide but they get close, causing an increased density in the areas of near collision. We see a similar effect in traffic with "rubber-necking" where a person slows down to look at something and then those behind are forced to slam on their brakes to avoid collision. If you watched this from above it would be similar to in traffic, which is a lot less complex than when it is occurring in a flock of birds.

We also see something similar in space and in galaxies. As the pre-matter "heaven" and "earth" of the first sentence of Genesis One were in a "waters" state, a catalyst, the firmament, was added into the mix and the "waters under" began to coalesce into substance as what we likely recognize as subatomic matter today. Then what we now call "gravity" would have been able to begin to draw the substance together. As the substance pulled together it would have formed increasing size clusters of matter until there was little or no free matter left to gather. This would have caused many random clusters of matter to form throughout the firmament of Heaven.

These clusters, now being greatly influenced by the gravity of each other, could then draw nearer to other clusters that had the greatest pull. In doing so they would be set and reset in motion having momentum. This effect would have repeated until everything was largely in balance, and we can assume that these clusters of matter would continue to move through and beyond full balance being achieved.

As the clusters of matter grow in size from gathering with other clusters they would eventually cause more and more intense collisions. This appears to still be occurring to this very day.

With all of these movements occurring, we can expect exactly the type of random murmurations that we see today in the form of randomly shaped galaxies with varying densities within them and varying distances between galaxies. We would expect them to be much like finger prints where they are all sort of the same yet they are all unique.

The gathering due to the firmament command "god said let there be a firmament made amidst the waters", would have occurred first at the subatomic level and then all the way through to the galactic level and beyond.

Some mathematicians have problems with this because they theorize an initial invisible fog of pre-matter would have been perfectly even and therefore the gravitational pull would have been in perfect balance and thus nothing could have ever moved. And even if it could have broken loose, then everything would be identical in size and spacing mathematically. Math is great and is very useful for describing much of what we see in astrophysics, but assuming that everything was perfectly even and would have had equal pull no matter where you were, makes too many assumptions for any of it to even be consider as a viable possibility. There are many problems with a pre-matter fog that would have been *perfectly* balanced, and *nothing* at all in our real-world experience supports the idea of a "perfect" balance whatsoever.

Chapter 11

Accept the Facts

We have to see the *facts* for what they are, and *speculations* for what they are. If the pre-matter was perfectly consistent as it may or may not have actually been, it is not relevant because it could have been in a perfectly consistent state for trillions of years and it would not matter a bit in regard to Creation as we see it today.

It was when the *be-ing* of light began that the Creative journey of what we see today as substance was initiated. And then it was the moment that the firmament began to catalyze the "waters under" thus dividing them from the "waters above", at which point math can really only be considered to begin to be used. It is the firmament that we must look to regarding our mathematics of the initial steps of Creation.

The Meaning of Our Math

Our science math is consistent and works because of the reliable constancy of matter. Math explains nothing and is nothing. Numbers do not exist because they are not something. One plus one equals two is misunderstood, as is two and two equals four. Two and two can be five if we make it five, but only in symbol alone and not in a physical or in a conceptual manner. I believe it was Shakespeare that wrote "A rose by any other name would smell as sweet", meaning that no matter what you choose to call it—it is still the same. Thus, if we make the symbol for five (5) to have the value normally assigned to our symbol of four (4) then 5 equals this many >I I I I< and thus it is only a symbolic gesture of representation and we have changed the symbol for the value >I I I I< which is the first point to understand about our math. In other words, the symbol used is a different attribute than the value we assign to it, for any given number in our ten-digit set.

The next point about calculating matter with math is that *matter* is extremely consistent to an astonishing degree. If it was not, then our math would fail with every astrophysical equation we attempt to make. Math works better than our human languages for communication because the various symbols used in the various languages all share one common attribute, which is that the value of a given symbol is consistent within any one language. And further, those values are interchangeable between languages even though the symbols may differ, provided that we agree on which symbols are equal between languages. It is our mathematical *consistency* that allows us to even begin measuring the unparalleled consistency of *matter*.

The Procession of Cycles

The procession of cycles, or "one great year", which is said to be 26,000 years, assumes that nothing has changed and that all things functioned over time as they do today.

The idea of counting vast amounts of time becomes problematic due to what was discussed in a previous chapter having to do with red shift light bending as it is being altered by gravity. But we also must consider in a historical Biblical sense that the terms *day* and *year* might not fully match what we experience today as a *day* and a *year* on our planet Earth.

Now in this case, I am not referring to on the first "day" or the second "day" but rather I am looking at the fourth day where "god said let there be lights made in the firmament of heaven to divide the day and the night and let them be for signs and for seasons and for days and years". In this fourth day text, it seems quite obvious that the light marking signs and season and for days and years is getting very close to what we experience today regarding those divisions. However, on a scope-of-view basis these particular mentioned "for signs and for seasons and for days and years" are very likely not specifically referring to our specific planet Earth's time-frame alone. Since at this point in Creation of the fourth day, we really do not know if it is speaking specifically about our planet Earth that we humans live on, or if it is that the "for signs and for seasons and for days and years" is all inclusive throughout *all* of space.

What is the "day" in the for "days and years" part of the fourth day? It very well may be the length of time that it takes for Earth to spin one full rotation and for it to travel around our Sun one full circle. But at this point, we again must ask, what is "Earth"? Is Earth only our planet? Or is "Earth" all planets? Could every single star or sun possibly have an Earth-like planet revolving around its star as the planet rotates its "day"? The stars would likely not yet have been fully formed at least until day four, so they might be included in the initial dividing of the "waters *under*" on day three.

Remember, the waters under were separated from the waters above. And then the waters under were later "gathered together" to make dry land appear, and that dry land was named "Earth". This in no way indicates that it is referring only to our planet alone. In fact, from the beginning and through and including the fifth day, the Creation account is stated quite generically in that it is referring to all of Creation and not just our little chunk of dirt we affectionately call planet "Earth". Further, the "Waters" that were "gathered together" to "make dry land <u>appear</u>" were still not necessarily the H^2O that we know as water today until they were called "seas" and the "dry land" *appeared.* This is a particularly important point when trying to understand the text. Since all credible Bible versions use the terms *appear* and *gather,* it is interesting that the "waters were gathered into one place". Without stretching things at all, if we stick to the idea that "waters" means the ability to flow then this makes complete sense. When the "waters were gathered into one place", those "waters" were still likely not the waters we see as H^2O until they were called the "Seas", at which point the land was no longer in a state of any kind of fluidity, but the Seas were, and still today are fluid.

On "days" one, two, and three there was no marking of days and years because that only occurred from day four onward. So here we are not discussing the type of "day" used when referencing the first three days. We are now talking about an issue of calibration, or the time it takes for any one planet to make one full 360-degree rotation, or that planet's "day". This could vary greatly from planet to planet as we know it does with planets in our own solar system.

Additionally, the same is true of a "year". *Year* lengths vary from planet to planet. Each planet revolves around its sun/star at a different rate and these rates are not specifically important. If a planet spins at a certain rate and makes the trip around its star at a certain speed then a given amount of spins will have occurred during that full complete circular trip around the star. For the planet making that trip, the spin count is the quantity of days or spins in its year. For our planet Earth it is 365 1/4 spins per year, which is why every four years we add a day at the end of the month of February which is referred to as a "leap year".

Other planets can have some random number of spins such as 784 1/7 and thus have a "leap year" every seven of that planet's years in that case.

"Processional cycles" of 26,000 years is the time we assume that it takes our Earth's axis to wobble back to any position it is currently in. Earth is often described as wobbling as it spins like a child's toy top. When you spin a toy top it tends to wobble, and the axis of spin makes circular or wobbling motions. This is what we believe we see with our Earth. It is believed that each full circle that the axis makes takes 26,000 years to complete. Is this relevant in anyway? Not really, other than it has been understood for a very long time that something causing long-term cycles might be occurring. This is good to know though, because historically, heavenly bodies have had their positions quite accurately recorded by us humans. Our understanding that this wobble possibly occurs helps us to understand specifically when something might have occurred in history from a previous ancient recoding of position of a heavenly body near the time of the recorded event. Yet, we really do not know if this supposed wobble has been constant, so again, our extrapolations are only our best guesses based upon what we see occurring today. As of the writing of this book, the only thing we have in order to make such assumptions is the apparent rotation and angle change of Earth's axis as recorded thus far. Beyond that it is guess work.

The Makers of the Meter made Their Intended Point

Hundreds of years before this book was written there was little consistency in measure. The term "foot" may have been used somewhat universally, however the actual length of a foot while it was called "foot" often varied from city to city. It's a real problem in commerce when measure doesn't match. If you need some strong steel cable and you need exactly 75 feet of it and you measure that distance with your tape measure, and then you order 75 feet from a store in our modern world, it will be the length you ordered and expected. But it wasn't always that way.

In the past, measurements could vary, where your measurement might have been 75 feet, but the merchant who sold you the cable might have also measured 75 feet very accurately, but his "foot" length was a little bit shorter than your measure of a foot. In this case you end up with only 70 feet of cable which is not enough to do the job because it cannot span the 75-foot distance you needed it for.

Is that merchant a scoundrel who cheated you? Not likely. Hundreds of years back before we had mass communication, mass travel, and firm standards, there were differences in calibration of measure from town to town. When doing business within any one town, the measure would work fine, but when doing business with a distant city, that city may have had a slightly different length of a "foot".

To remedy these discrepancies a couple of men during the French Revolution took up the task of trying to create a consistent unit of measure that would stand the test of time. These men attempted to measure the circumference of the Earth, or a quarter of it, and the "meter" was to be 1/10,000,000 of that distance because they assumed that this distance would not ever change. This way we could always check the meter's calibration against the Earth's circumference. Was that a good idea or a foolish idea?

It was largely an unnecessary idea. It was certainly a noble effort on their part, but it still won't solve the problem of differing lengths of measure. What merchant has the time to measure the world after all? As mentioned earlier, regarding the value we give to our numeric symbols, we can make a 5 to be a 4 if we want to, and so long as we agree on the true accurate value of the symbol, it simply doesn't matter how the symbol looks.

The same is true of the measure of distance. A foot is a symbol of distance, and as long as the foot I measure with and the foot that you measure with are identical in actual physical length then all is well. But when our symbols for a foot, which is a stick of a given length, do not represent identical lengths, it would be like one person using 4 = >I I I I< but another person using 4 = >I I I I I< so one person using 4 expects >I I I I I< but only gets >I I I I<. What the Frenchmen achieved in making the meter was simply a standard distance agreed upon and to be used by everyone in the country. The increments and length of that standard are meaningless provided they stay consistent and are agreed upon and used by all. This is clearly illustrated today by the fact that we use a standard calculation with high accuracy when converting standard measurements to metric measurements. It is always the *index value* that matters. Agreed/stable/unchanging values are what make things possible, and are why the contents of the Heavens exist today. *Standard values* are what Genesis One establishes throughout.

Point Proven in the Errors of the Makers of the Meter

A consistent unit of measure is really all that the makers of the meter wanted to achieve. A meter is a useless term and a useless value unless we all agree on its exact value. There was no need to measure the Earth, as noble as that cause was. Their measurements were actually somewhat inaccurate. A minor discrepancy it is true, but an error nonetheless. Because the error existed in their actual measurements we are now stuck with a meter that is not the intended fraction of the earth that they desired to achieve.

Does this mean that the meter is all wrong? No it does not, what it means is that they wasted their time in this regard. They were after a firm and reliable index and thought that they had found it in the Earth's size, yet all that really needed to happen was to create grand standards that everyone anywhere on earth could use as references for size, weight, and temperature. Let us not forget that there also was a push away from the old system to begin anew during the French Revolution, which is maybe more of a contributor to the stepping away from the measure of an imperial "foot" than anything else.

The point I am making is that the unit of measure is irrelevant as long as we all agree as to what the true and actual value of a particular unit is. Even though there was a mistake in the measurement of Earth's circumference, the meter still worked well as it does today *because we all agree on its value.* For many people, the unit "foot" or "feet" still works just as well as the meter and will continue to do so as long as those values remain constant.

Most of the issues regarding the particular indices used in measure have to do with honest discrepancies. However, people did cheat back then just as some do today by short-changing people in measure, albeit relatively rare today. It did not matter what the meter was, as long as there were rules or laws to largely abolish honest discrepancies and/or cheating with dishonest measure.

Cheating Still Goes On

Cheating still goes on, though in far less profound a manner as it may have in 1700's France. Instead of dictating the size of a loaf of bread, now sellers and manufacturers are required to state its price basis and or weight on the package, or display price nearby, thus allowing *us*, the consumers, to decide if it is a fair deal rather than deceiving us as in the days of old with inconsistent weights.

There is another type of cheating with weights, but it is not really cheating as such. When we look at the cost of "generic" items in the store, where we often see that we can buy identical items in much larger quantities at a similar price, but in this case, we get to choose this as we grab an item from the store's shelf. Generic items have little or no advertising overhead to recover in the price of the items, which is one of the biggest reasons that generics cost less. So, it is not really cheating in that case. Where we see the most cheating today might surprise many who read this. There is virtually no field or industry that cheats measurement more than in sciences of astrophysics. The ranges given for distances and for ages often will vary considerably from one scientist to the next. Some of this is due to the difficulty of the task such as when the two Frenchmen decided to measure the Earth's circumference. Somethings are simply very difficult to do and all we can do is to do our best.

But we also have those who will bend that ruler to fit their version of the "facts" so that the numbers that *they* want to be true appear as if they actually work; most notably, in anything having to do with the big bang and all of its connected calculations. If merchants did this same thing with bread today they would be fined and or jailed for their dishonesty.

Chapter 12

Our Measuring Tape

Our measuring indices are very important if we ever hope to continue to advance scientifically or even advance in our understanding of anything. When we trap ourselves in erred ideas then our advancement becomes impeded.

Measuring the Meridian

As I mentioned in the last chapter, two Frenchmen measured the meridian of Earth. Their names were Mechain and Delambre. They embarked on this mission to measure the meridian through the method of meticulous triangulation. Their goal was to create a standard of measure that had as its index our planet Earth and it did succeed... to an extent.

In trying to create the metric system they went so far as to change the calendar and the clock, but this failed miserably due to its rejection. Why did it fail? Because we already have a standard established and it is the single rotation of the Earth from zenith to zenith. It does not matter what we call the increments that we use to measure this rotation, nor does it matter how big the increments are. What matters is the duration from zenith to zenith, which is something we have no control over, thank God! Humans cannot cheat a day by alteration in any way. There is but one day on Earth, and everything that we do is based upon zenith to zenith rotation, or a single complete 360 degree turn of our Earth.

When we make measurements in science there is an incorrect perception that we are very accurate. This was partly discussed in an earlier chapter where when we divide a scale of measure into very fine increments, it is not any more accurate than the measure itself. It is higher resolution for sure, but that has little relevance regarding accuracy in the bigger picture. If we take a mechanical military twenty-four-hour clock and we divide that military clock into one million divisions, and then proceed to over-excite the motor to speed up the clock, then the entire twenty-four-hour span is incorrect. The divisions on the clock are nothing more than one-millionth divisions of the inaccurate twenty-four-hour military clock in that case. It would be a very high-resolution clock, but now one of its million increments is still not a millionth of a rotation of planet Earth in that case.

This is important for you to understand because when scientists talk about things like astrophysics, gravity, time, space, distances, atomic clocks, and high-resolution methods of measure, it is often inferred that the measurements are accurate, giving us an erred impression of overall accuracy, which is simply not true. It's not that they're inaccurate, but rather high-precision does *not* equal high-accuracy.

The Mile High Ruler

When trying to grasp accuracy, consider this visual. I want you to think of a distance of one mile that you are familiar with, maybe down a country road or several city blocks. Now imagine that distance as a single stack of paper standing upright with one piece of paper lying flat upon the next piece and each piece of paper being only about .004 of an inch thick (4/1000). This is about the thickness of a typical piece of paper you might put in your computer printer on your desk, or a piece of typical typewriter paper if you are familiar with that. When doing the math, you will find that at this thickness of paper, one inch of paper on the stack will be 250 sheets. One foot will be 3000 sheets and a stack a mile high will have 15,840,000 sheets.

So as to not miss any zeros, that's nearly sixteen million sheets in the paper stack. A billion sheets would be a stack about 63.13 miles tall. 15.8 <u>b</u>illion sheets would be one thousand miles high. 1000 miles is a lot of miles to picture, so we will use one mile as an index.

Now, if we multiply 15.8 <u>m</u>illion by one thousand we get the number 15.8 billion with a "B". This means that each piece of paper in that mile-high stack represents a value of 1000. For this point to be grasped, the 1000 value for each paper could be any unit of measure—It can be time or distance or anything else for illustrative purposes.

Now let's take one single piece of the paper from the top of the mile-high stack, because we wouldn't be able to lift the stack to pull a piece out from the bottom, because a paper stack one mile high would weigh roughly sixty-five tons depending upon paper density.

Now that we have this piece of paper we have to realize that a single sheet thickness represents roughly 1000 years in the suggested age of the Universe. If we scale it to our own life experience, we must use the average lifespan of a human or roughly 77 years, but we will use 100 years for simplicity sake. To represent a human lifetime of experiences relative to the suggested age of the Universe we will have to slice the thickness of that single sheet of paper into ten equal layers. It is one of those 1/10 parts or layers that is representative of your *entire* lifespan within the speculated lifespan of the Universe.

What do we have that can illustrate that thin layer that is one tenth the thickness of a sheet of paper? Not Bible paper or tissue paper, those are about half the thickness of our original paper. The thickness of a typical single layer of cellophane or plastic wrap is roughly your *entire* life span in that one-mile-tall stack of paper. Feeling insignificant yet?

The point of this illustration is for you to get a picture of scale when astrophysicists talk *billions* of years. The one-mile high stack of paper approximates the assumed age of the visible Universe at 1000 years per piece of paper, and your plastic-wrap thickness one tenth the thickness of a piece of common paper is your **entire** lifespan in comparison to the span of time represented by the entire one-mile-high paper stack.

Average versus Normal

In a non-science, but Biblical reference in Revelation it says "Hold fast that which thou hast, that no man take thy crown". We are probably reasonably safe in assuming that the "crown" spoken of here is not a king's physical item of a crown. The "Crown" is likely that which makes them special, whole, or saved. We must learn to hold fast to true wisdom and understand data for what it is.

When we allow erred theories into our thinking, we cast away our crowns and trade them for the hat of the foolish court jester or worse, we trade our crown for a "dunce hat". This does not mean that we should ignore theories that do not agree with our own thinking. What it means is that we must examine with open mind, any theory that is not in agreement with our own. We then use truth to discern what of the theory, if anything, is realistic.

In our modern culture with our vast network of communication devices, we can distribute our words at a rate which is unparalleled in history. With this tool we can distribute deceptive or incorrect information very quickly to a multitude of unsuspecting people. We see this with the news outlets on a regular basis, where scientific *opinion* is presented as *hard fact* when in reality it is not fact at all. These deceptive practices also occur in religion and in science.

For instance, when someone presents an "average" and refers to it as "normal" it is potentially a very dishonest statistical figure. Consider this example: If there are ten people holding cash in their hands and five have nine dollars and four have eleven dollars and one person has one-thousand-eleven dollars, then the normal would be close to ten dollars per person where the "norm" or normal that *most* people have is roughly ten dollars and the person with one-thousand-eleven is not in the same category as the others and is therefore not "normal" as the masses are, and should be considered to be an outlier. But the "average" in our example is that everyone has one-hundred and ten dollars. This would likely make the people that had nine or eleven dollars in hand quite happy, but it is not reality.

Often when people do misapply such terms it is to present information in a dumbed down language so that it is more relatable for people who don't really care about it all that much. Weather forecasts are a good example of this. It's very common to hear the weather forecast presented as "the *normal* temperature for this time of year is XX degrees." What they really should always say is the "*average* temperature" because they are most likely giving the figures of the averages when they present that information to us. No big deal and no harm done when there are no major spikes in the historical data. And besides, it's only weather temperatures.

But when we look at the issue of "global warming" or "climate change" then it is a bit more important to be accurate about the information and the way in which it is presented, because the claim is that a very minor temperature change in the oceans will cause them to overflow and we will lose a great deal of land-mass as the shores are consumed by the oceans. Is this true? Only time will tell. Since the fears associated with climate change have been scattered throughout the people, many weather reports claim "highest ever recorded temperatures" or "hottest summer on record". This all may be true, but there are many factors to consider in understanding those "record-setting" statistics.

The first and most obvious factor is that the term "on record" is a very short time period. Remember the one-mile tall stack of paper mentioned in the previous section? Just two of your entire life-spans of one thin plastic-wrap film's thicknesses or 1/5 the thickness of a typical copier paper is more than the amount of the length of "on record" that is typically referred to by sensationalized news reports referencing "record" temperatures. In fact, many of the on record temps go back less than one hundred years depending upon news outlet.

Additionally, we all have to understand that temperatures are very tightly regional. Many cities in a region will have differing temperatures on any one day. And just one of those cities might have reached a new high temperature, while neighboring cites did not reach a new high for their city at that time. But only the record city will be reported on the news because it is a new "record high" that will help them to embellish their headlines. And it is wise to take note that not all cities began recording records at the same time. Often a record will state the old record date that the new record beat, however, while one city might have been keeping records for over a hundred years, a neighboring city might have only been recording for the past couple of decades. We all must realize that there is a bit more to records and statistics than meets the eye.

Now add to that the fact that often the "record high" is typically for a specific day of the year such as August 5th where only the temperature numbers for August 5th for every year recorded are being considered in the particular report. But the previous day and the following day of the record day could be substantially lower or higher. If you watch for this you will notice it on a regular basis when listening to these sensationalized weather and climate reports.

This is typically very dishonest reporting for nothing more than the attention-grabbing sensationalized headline that it is. Most people who have lived more than fifty years are going to have a good idea of what I am referring to because they are old enough to have experienced much greater weather fluctuation during their lifetime than younger people have thus far. A twenty-five-year-old person will likely not yet have experienced the same extreme range of temperature highs and lows as a fifty-year-old person. If this is not clear enough for you then find a stock market chart and consider the spikes you see as you zoom into a narrower and narrower time-frame of any given chart. Doing so will illustrate the importance of *sample size*, which is among the most abused reference points in science, especially in astrophysics.

Everything moves and fluctuates and we must understand that while we theorize using *any* measurements or calibration. Sample size is a very critical part of estimating in *every* scientific field. And accuracy becomes ever more unstable as the sample size is reduced.

An Astronomical Unit

With regard to this book, a key measuring method is a lightyear. We also have an "astronomical unit" which is the distance between our Earth's center and our Sun's center. When we refer to such units it is important to understand that these are indexes of approximation only.

While a lightyear would theoretically be a bit more precise than an astronomical-unit, it is in truth only a mathematical theoretical figure, but could ultimately vary in length in real-world measure. A lightyear is a vast distance that no one has yet to walk.

An astronomical unit is a bit more concrete in that it is the distance from Earth to Sun, but it too has the problem of fluctuation as the Earth orbits the Sun. As Earth travels around the Sun, the distance fluctuates roughly three million miles in distance or just over three percent. Three percent of a 12 inch ruler is about 3/8 of an inch—It's a big difference.

How Many Times?

A key area of volumetric estimation that often eludes our minds is order-of-magnitude. Think in terms of ten times the size, or a hundred times the size.

If you are given a one-inch cube of steel and then are given another, then do you have twice as much cube? Yes, you do. Now if you were to combine those two cubes into one big cube then is it still twice as much cube? Yes and no.

If a cube is twice as big then its measurement would be two inches cubed or a volume of space equal to eight of the original one-inch cube. Some of this comes down to semantics, and it is important for you to understand that those basic semantics exist and that they are often abused in astrophysics.

We must all understand that we might not be understanding something as someone else is describing it to us. Is a pile of dirt that is twice the size of another pile ten feet high rather than five feet high? Or is it just two of the five-foot-high piles of dirt combined into a single pile? The amount of dirt in those two examples is *very* different.

We must also consider that *weight measurement* and *volume measurement* can be very different. When done in ounces for instance, eight volume/fluid ounces of feathers in an 8oz cup is far lighter in weight than eight volume/fluid ounces of water in an 8oz cup.

If something is X times the size of something else, there can be more than one interpretation of that X value. We all need to understand that we might not be understanding something as someone else is describing it. If someone says something is 30 times more, then some people might see it as 30 times the weight, or 30 times the volume or 30 times the dimensions. Now weight and volume will closely coincide on a substance such as water, however 30 times the dimensions will be far greater when weighed than 30 times the weight will be when weighed.

To clear this up a bit, 1 x 1 x 1 = 1 and 30 x 1 x 1 = 30. However, 30 times the length, width, and height dimensions is 30 x 30 x 30 = 27,000. So, when we listen to someone saying that something is "30 times" we either need to think very cautiously about their words, or ask them to be more specific.

Saying that something is 30 times the mass does not necessarily have to be where it had a mass of 10 it now has a mass of 300. A discussion regarding "mass" can become a bit ambiguous very quickly because we do not fully understand gravity. And the "Mass" of something is somewhat arbitrary. Mass is only our index of stating the measurement we believe something to be, and it has been working as a good and reliable index for a long time.

But when it comes to celestial bodies, we can only speculate. For instance, if a planet is 10 times the mass of Earth then how can we tell this? In order to do so we must be certain that we understand which elements that planet is made of. When the exact composition and quantity of elements are unknown, then we are somewhat guessing. Now of course we can get a better idea by looking at its size, speed, and its distance from the its star. But this still does not reveal everything about the planet; it is still all just our best scientific guesses.

We could land a probe on a planet and assess the surface material, and then make a better estimate of its composition and mass from that data as well, but since we are only literally scratching the surface, it is still nothing more than a calculated guess. This does not mean that we should not make such guesses, what it means is that there is much more to learn. And we should continue to work to understand our stellar environment rather than making concrete assumptions as far too many pop-scientists do.

Unit of Measure is Not Important, but a Single Index Is

Many people never consider these things because science is not their field of expertise. Yet they will set their entire belief system upon misunderstood data presented by pop-science. The units of measure are not important, but a single index for each type of measure is very important in our efforts to advance our understanding of the heavens. The only thing that is important is for all the world to agree on the same indices and the values for those indices, along with the relationship between the symbols that represent the numbers from one language to another language.

When scientists cheat on the measure or magnitude, or if we misunderstand the measure or magnitude, then we can all arrive at very different conclusions, which is something we see quite often in pop-science. If we deeply "believe" our erred understanding, then we tend to fight evermore passionately about our erred conclusions that we so deeply believe regarding the particular topic. When you consider that truth is an abstract concept in the minds of some scientists, you can understand why scientific explanation gets so convoluted. Truth is not abstract in any way. *Truth* is the only perfect standard!

Chapter 13

Our Changing Minds

Our changing minds typically do so reluctantly. This protective human attribute can be our downfall when we fail to properly discern any new data that we receive. It is good to not suddenly shift from one belief to the next, but it is very bad to not release erred beliefs. When we suddenly change our thinking, we should be very cautious to first carefully analyze that data that convinced us to change our mind. Deep down, most people know when they are being duped, but there is so much informational noise out there for us all to consider that, all too often, our better judgement is drowned out by this noise, often causing us to completely disengage.

Information flows like water, and when that information is incorrect that water is dirty, yet it still flows. If you drink contaminated water you will eventually become infected or sick. It's no different with the scientific information you choose to believe. If you filter your information water through proper discernment then it will be clean enough to support life, thus allowing you to continue in your scientific understanding journey.

Failing to do so can cause your thoughts to become infected with filthy incorrect information which will cause you to draw incorrect conclusions.

Having Reservations

Our current world-view of history is often obscured by "experts" who have drawn conclusions for us. Throughout history, people have done what many of us do today where we trust the "experts", thus expecting that they only tell us the truth. But all too often what they claim to be the "truth" is incorrect. This occurs so readily and so repeatedly because most of us are honest and would never deliberately intend to lead someone down an incorrect path, and so we assume the same level of honesty from others.

Throughout history, people who might have had a vague interest in something like how the stars and planets came to be, generally trusted those who studied the heavens. They simply did not know anything about the subject, and the information from the experts was good enough for them. Some people will believe "experts" who show deep conviction. But those who have greater wisdom will reserve their conclusions due to lack of confirmation of the information that they have been given from such experts.

We can group the entire populous into several groups:

- Those who saw and understood.
- Those who created myth from misinterpreting evidence.
- Those who simply did not care.
- Those who followed myth.
- Those who followed what appeared most logical.
- Those who simply did not notice, care, or even think to question.
- Those who have reservations about the information and cautiously wait to make final conclusions.

Regardless of which group you yourself might fall into, your belief style from above, affects your life and the lives of those all around you. Let us not become like those who buy into myth or those who create myth.

The Bible is often thought of as "myth" by atheists, yet the Bible has more concrete evidence of people, places, and things than little other ancient history has ever come anywhere near to having. On the other hand, the myth of the big bang has little or no supporting evidence. And the claimed "evidence" that has been found regarding the big bang is not necessarily the result of a big bang.

But perhaps even more disturbing is when Christians or believers fight for the utterly absurd notion of six twenty-four-hour day creation and insist that it occurred that way when their frame of reference is skewed and their explanations do not match the actual text or real-world findings and facts.

Quantum Entanglement

Quantum entanglement sounds like serious stuff, and when you get into the deeper details, it is a complex experiment. But for our purposes here we will keep it simple.

Quantum entanglement is a thus far unexplained connection between two paired particles. When the particles are separated or split, then the two are paired. When one is in one state or position then the other is claimed to be in an opposite position.

You could think of this in terms of the three-way light switches many people have in their homes. When one switch goes up then the light goes on, but when the second switch goes up then the light goes off. Now when the first switch goes down the light goes on again thus changing the state of that switch. The off/on positions of each switch change according to the paired switch on the other end of the circuit. But with quantum entanglement there are no wires connecting the particles together and the particles are not turning on and off but rather are changing in charge state of positive or negative as stated.

In theory, quantum entanglement occurs instantly at any distance apart that the particles are. There currently is little understood about the inner workings of this "entanglement" effect and all that we really know is that it seems to repeatably exist.

But the interesting thing about quantum entanglement is the supposed instantaneous nature of the action. If this is in fact occurring as is claimed, then a Creator of such particles would be able to command, Create, and communicate instantaneously throughout the entire universe, and the notion of an expanding, growing universe could then be nullified or rendered unnecessary for Creation to occur. This would mean that Creation could instantly begin everywhere simultaneously.

I am not suggesting that everything was made instantly, but rather that Creation's first point of entry could have been everywhere at the same time when it was commanded to **Be**–as in "be light made".

Initial light substance did not have to spread out at the speed of light throughout space taking billions of years to get from here to there. Rather, the possibility is that the matter that light consists of was initiated throughout the entirety of "heaven" simultaneously. This is **not** suggesting that light instantly would be seen from stars that are light years away, but rather that it is important for us to understand that something could occur simultaneously throughout all of Creation. We have a tendency to visualize things occurring only the way *we* see things occur in this physical world, much like if you poured out water on the pavement and the wet area would grow as the water seeps its way across the pavement.

But with the idea of quantum entanglement, we can see in a physical sense that it could theoretically occur everywhere instantly, which would be like pouring the water out, but the moment the first molecule of water hit the pavement the water would all be instantly the entire size of the wetted area that a normally poured out amount of water would have been able to flow to its full and final extents.

Additionally, we have to realize that anything in the quantum realm is still Created matter. And since the Creator would not have been Created as such, a similar action could have occurred with what I have previously referred to as pre-matter or pre-substance. This is not a point I care to get to deeply into because it steps outside of simple logic that we have been using discussing the first four days. It is only mentioned here as an additional interesting point. To base any conclusions on the idea of quantum entanglement or anything similar would be purely speculative and is not specifically mentioned in Genesis One.

Christians Should Not Leave Reason at the Door—Ever!

The twenty-first-century science world is often thought of as using only *reason* with which to draw their conclusions. As a Christian you should not leave reason at the door when reading your Bible or attending church. It is actually quite the contrary. We should test it all if we are Christians and it must be harmonic with the Bible or we have a problem in either science or in our understanding of the Bible. Whatever occurred, occurred, and it is our goal to uncover what actually occurred. Will you use your Bible as a springboard for understanding science, or will you use it as chains to hold you in bondage? Interpretation matters!

If a true Creator actually does exist, do we then imagine that that Creator who is supposed to be "perfect" would want us to cheat or lie to ourselves and others in order to make our interpretation of Genesis One make sense to us and others? Even the thought of that is irrational if the God in the Bible does in fact exist.

Believers, and especially Christians, are more culpable when they make errors in their beliefs or statements about Creation than are atheists. If people who claim to be "God's own people" fail in this regard, then what is the rest of the world to think?

Believers have a much greater scientific task than do big bang believers because big bang believers can invent any theory no matter how absurd, and since many do not believe in a discerning Creator, they in their own minds have nothing to fear. But with Christians they face hell-fire when they lead others down foolish and dangerous paths. *Never* leave reason at the door, and always *discern* the information you receive no matter how scientifically certain it might sound to you at that moment. Do not become a false preacher of lies, especially with the Creation account of Genesis One.

Who decides what is Scientifically Acceptable?

What is acceptable scientifically is decided by people who do science and believe certain things. This information is then broadcast to the world and is often accepted as "fact" by gullible teachers of young people. The young people range in ages of about four years old to about twenty-five years old and most have not yet learned to properly analyze and discern the information that they receive, nor are they taught to do so. In many cases, they are told what to believe, rather than being instructed to look at the evidence and weigh it against the various theories that have been proposed to them, or even think up, on their own, a theory when the data they receive doesn't sound correct.

Often, the decisions regarding the *acceptability* of scientific data are more accidental than they are anything else. This occurs because a person of science will have a new or altered theory that is newsworthy. Then an interview is held or a paper is written and the theory is introduced to the world. It is then picked up by the gullible teachers and repeated to the undiscerning students, and far too often it is held as a *belief of fact* rather than as scientific *theory* by those teachers. It is then subsequently taught as "fact" to the students.

If You Believe Lies You Have to be Evasive

When you look, you will find a reliable pattern where people who believe erred information will be evasive when they debate a topic. They can be the foremost expert on the theory, but if some of their conclusions contain anomalies, then they will be forced to fill in the gaps or evade the flawed aspects of their theory. As mentioned in an earlier chapter, when believers do this it is referred to as "the god of the gaps". Yet somehow, the scientific community and many other people overlook this when big bang believers do it on a constant basis.

When studying any subject, you should only draw conclusions that can be drawn without compromise; everything else should be held in consideration until *uncompromised* conclusions can be drawn.

Whether you are a scientist, or a believer, you should never have to be evasive and you should never have to say "The Bible says so", or "Hubble or Einstein said so". Data should always be able to stand alone and hold up against all reasonable, proper, and fair scientific scrutiny; or we should simply admit that we are not certain about a particular point.

Chapter 14

Is There Life Out There?

This book is promoted as being about the first four days of Creation and it is. But since Creation occurred in "day" groupings it forces some mention of plant life. Day three is discussed in-depth in *The Science Of God - Volume 2 - Day Three - Gravity, Land, Seas, and Evolution of Plants,* but is only briefly mentioned in this volume which is intended to be about the astrophysics aspects of Genesis One.

Life as we understand it today could not have existed until the at least the third day, but some say the fourth day. The text reads:

"god also said let the waters that are under the heaven be gathered together into one place and let the dry land appear and it was so done and god called the dry land earth and the gathering together of the waters he called seas and god saw that it was good and he said let the earth bring forth the green herb and such as may seed and the fruit tree yielding fruit after its kind which may have seed in itself upon the earth and it was so done and the earth brought forth the green herb and such as yieldeth seed according to its kind and the tree that beareth fruit having seed each one according to its kind and god saw that it was good and the evening and the morning were the third day"

And then:

"and god said Let there be lights made in the firmament of heaven to divide the day and the night and let them be for signs and for seasons and for days and years to shine in the firmament of heaven and to give light upon the earth and it was so done and god made two great lights a greater light to rule the day and a lesser light to rule the night and the stars and he set them in the firmament of heaven to shine upon the earth and to rule the day and the night and to divide the light and the darkness and god saw that it was good and the evening and morning were the fourth day"

From a biological perspective, there are many people who believe that nothing could grow because the Sun was not in place, or at least not shining until the fourth day. As the text reads, it is true that the Sun could not have cast its light upon the Earth until the fourth day. However, Sunshine is not a prerequisite for all of life. A plant might not be able to continue its growth very long without sunlight, but that does not mean that it cannot *begin* without sunlight.

Consider a plant. Can a plant germinate without sunlight? The obvious answer is yes. In fact, many seeds will die if they are unable to root into soil that is free of sunlight beneath the soil. Seeds can sprout in the dirt under large rocks that completely shield the seed from the sunlight. If this is not evidence enough, then let us consider the living plants found in the right conditions on the ocean floor. There are areas deep within the oceans where life flourishes absent of any sunlight and likely receives no radiation from the Sun whatsoever. So, it appears that from our real-world experience, we can easily determine that the sunlight is not a prerequisite for *all* stages of bio-activity.

Life Hangs in the Balance

What exactly can we realistically deduce from the third day's Creation events described in Genesis One? First, we have to get a grasp on the scope of the investigation. Did the seed suddenly sprout forth fully grown and bear seed after its kind, and the tree suddenly sprout up fully grown and bear seed after its kind?

Conventional scientific thinking about the Bible regarding six twenty-four-hour day creation would have the trees and plants fully grown in a single day. But here again we have this problem of cheating nature.

If the trees were Created on day three then how did they grow without any sunshine? Did the trees have growth rings? Were they big trees or seedlings? Were they some sort of super tree that could grow in the cold of space or in the heat of a newly created planet? Or did God create them whole as adult trees producing seed each after their kind? And further, is Earth's bringing forth of green herbs and fruit trees on day three exclusive to our planet Earth that *we* walk upon? There are many more such questions that can be asked in this regard, but let's see if we can answer or eliminate any of those thus far mentioned. Consider the text again:

"let the earth bring forth the green herb and such as may seed and the fruit tree yielding fruit after its kind which may have seed in itself upon the earth and it was so done and the earth brought forth the green herb and such as yieldeth seed according to its kind and the tree that beareth fruit having seed each one according to its kind"

As mentioned at the beginning of this chapter, for life to begin, it does not require the sunshine. Here you may infer some contradiction, but there is none. When a seed germinates, it does not specifically need light from a sun. I am not referring to artificial light either when indicating that sunlight is *not* required. All that a seed needs is the proper conditions of moisture and heat and it will be able to germinate.

Since the water under the firmament was gathered and called the "seas" on day three and dry land appeared on the same said day, most of the required conditions for germination were being met at that point. But we still have the matter of heat. Since the two great lights have not yet been made in the firmament there would have been no heat. Or would there have?

As we understand heat today, heat is a result or a part of the electromagnetic spectrum just below the visible red part of the rainbow in the infrared area within the electromagnetic spectrum. The question here is what exactly is the "light" referring to during Creation's day one events? Without getting into the details of "light" too deeply here, there is a good possibility that the initialization of the electromagnetic spectrum, as we call it, was being established on day one. This would allow the possibility that "heat", as we call it, could occur given the right conditions on day three.

Now the day two events when the "waters under" the firmament were divided from the "waters above" the firmament, might not necessarily have created heat, though it is a real possibility. The elements might not yet have been formed as we understand them today in our periodic table, thus it is possible that no heat would yet have been produced on the initiation of the day two event.

However, on day three we can scientifically come to no other conclusion than that there was much activity that would have generated heat and potentially lots of it. The length of time of the day three events is unknowable or at least unknown to us. As our understanding of science and our own pure logic would dictate, the "waters under" would have to be separated and dry land would be required along with water of the seas on the "day" land was called "Earth" and it was the "Earth" that brought forth "the green herb and such as may seed and the fruit tree yielding fruit after its kind which may have seed in itself" Here we now have the conditions for potential bio-life. We have the water and earth materials as we know them today, and we have heat. While it was likely a very dark place, life could begin to form at that point. But to what extent could life form? And how narrow of a location is it referring to?

Being One with Nature

With the conditions set, life could begin, but how? And what is implied in the text regarding the various kinds of fruit trees and green herbs, and for that matter what specifically is "fruit"? See *The Science Of God Volume 2 - Day Three - Gravity, Land, Seas, and Evolution of Plants.*

The day three text does not dictate how many herbs or trees were brought forth, nor does it tell us how big they were or how long it took them to be brought forth. The text is very basic in that regard and only states that the events occurred.

This is where many Christians completely lose the argument. We are often told that God created all of the different trees that we have today and that they have always been the same. This is what we were told, or at least it is what too many of us think we were told. Yet the Bible does not say, nor do many and maybe *most* priests believe or teach this. Some preachers on the other hand are a bit of a different story in this regard, and some of them have a very biasedly-formed opinion on these issues. But here again there are many who are open in their thinking in this regard. Again, here is what the text says:

"and he said let the earth bring forth the green herb and such as may seed and the fruit tree yielding fruit after its kind which may have seed in itself upon the earth and it was so done and the earth brought forth the green herb and such as yieldeth seed according to its kind and the tree that beareth fruit having seed each one according to its kind"

Can you count the quantity of kinds in "according to its kind" for either "herb" or "tree"? No, you cannot because there are no statements of the number of kinds brought forth here for either the green herb or for the trees that bear fruit.

When we get into these areas that are a bit more complex and yet are too vague, I like to return to some of the older versions or previous languages that subsequent Bible translations may have originated from. But for brevity sake I have been using the

English Douay version of the Latin Vulgate version of the Bible, but it could just as well be the older King James version:

English Douay Genesis 1:12 "and the earth brought forth the green herb and such as yieldeth seed according to its kind and the tree that beareth fruit having seed each one according to its kind and god saw that it was good"

Latin Douay Genesis 1:12 "Et protulit terra herbam virentem, et facientem semen juxta genus suum, lignumque faciens fructum, et habens unumquodque sementem secundum speciem suam. Et vidit Deus quod esset bonum"

Other languages are shown here because we often get too accustomed to specific words that might be used in a particular favored version that *we like*, but the word might be wholly different than in another version. Often, antiquated versions done in languages unfamiliar to a person are the best to use when the tough questions arise regarding a particular word being used, this assists us in better understanding the original intention of that word when we take time to scrutinize the particular word in question.

The point is that day three is not likely referring to the plants as we know them today, and the Genesis One text makes no reference regarding anything to do with quantity of kinds other than that there were at least two types brought forth that were established, and they were the "green herb" and the "tree that beareth fruit". Yet we have this little part of the text "and such" which depending upon how the Latin is chosen to be interpreted, may or may not be accurate. If the "and such" was implied in the original or oldest text, then that implication would open the door for just about any vegetation interpretation we could dream up. So, we will ignore that potential in this book and stick only with the "green herb" and the "tree that beareth fruit" because they are all that is listed.

Again, how many kinds are listed? None, is the proper answer. There are two that we assume to be plant types listed, and that is the end of the text regarding those types. Or is it?

What is really occurring on day three? And what really does come first, the chicken or the egg? This puts us in a bit of a quandary in our narrow-minded scientific views. For a six twenty-four-hour-day Creationist there is no problem, because someone could just say God made seeds for all of the different plants that we have today and planted them with his hands. But that would be a fool's approach to the subject because of the obvious thing we humans are able to do to create differing breeds of plants, such as apple trees.

Scientifically we assume that all plants come from seeds, so to remedy this in science we evolve them from nothing and over billions or at least millions of years, thus in our minds we morph them into our current day plants. Both approaches are a bit of a stretch of the imagination when you dig deep into the component parts and the reality of nature and the inner working of the cells that make "nature" possible.

Day three indicates:

"and the earth brought forth the green herb and such as yieldeth seed according to its kind and the tree that beareth fruit having seed each one according to its kind and god saw that it was good"

"According to its kind" is a bit ambiguous here because it doesn't have a clear indication of what the "kind" specifically is referring to. This is once again a scope of view issue. Is the "yieldeth seed according to its kind" referring to the difference between the green herb and the fruit tree? Or is it relegating each to its type plant in that the green herbs would stay within the herb families and the fruit trees would stay within the fruit trees families and that any variation of green herbs or fruit trees would be locked in their own seed "kind" such as an apple tree, or a pear tree, or an orange tree, etc.

Is the reference to tree accurate, or is there a more clear possibility in the Latin version? Latin's "lignum" can translate as "stick" or "wood" or "tree" with each giving us a very different view in context. We also have the term "fruit" to deal with. What exactly is fruit? In Latin it is "fructum". This seems simple enough, it is where our word fructose comes from, commonly known as a type of sugar. But it is unlikely that there would have been any sizeable vegetation as we see today if the sunlight was not yet present.

Without getting into some very off-track language topics regarding this book, we will not go any further into understanding the underlying translations other than to take the Latin variations of "stick" or "wood" or "tree" and explain that tree is not necessarily referring specifically to the trees as we think of fully-grown trees today. "Stick" may very well be a better visualization of what was intended in the text, not that a stick was made or grew but rather a stick might more accurately represent the initial form of a young plant. This approach could bring a lot broader and possibly more accurate meaning to what actually occurred on the third day.

CO^2 Level Data

With all of our modern media sources, we often see charts showing skyrocketing CO^2 levels. These charts are created using information from examining core samples of ice and dirt. Each small section of the cored material is analyzed to estimate the amount of CO^2 or other molecules that may have been present at the time the layer being examined from the core sample was laid down and eventually buried. Then the data from those readings are plotted on a graph and the charts are presented to us in effort to give us a visualization of intensity of change in the concentration of an element or molecule in the various strata/layers of the analyzed core sample.

The problem we run into with this is, again, scope of view. Consider our one-mile high stack of paper and our super thin plastic wrap lifespan. All we can do, scientifically, is to take what occurs today and use that as our index and extrapolate that back as far as we can in relation to the core samples we bring up from the ice.

However, we really don't know *dissipation* or *decay* levels for certain, so a spike that we might find when testing air or materials from our current time could end up greatly exaggerated compared to the extracted core samples. If that same recent time sampling sits for a few hundred years the reading will likely be somewhat different. Additionally, the current-day spikes are not necessarily determined using the same methods. This is to say that the CO^2 readings for recent times are more likely to be open air samplings rather than samplings that were trapped in ice for some thousands of years. For anyone to expect two very different sampling methods to produce similar results is particularly disturbing and potentially very dishonest when these points are not plainly illustrated and made obvious when shared with readers as the visuals are presented. And even though they might be added as footnotes, we all know that most people do not care enough to study the footnotes, and seldom do news reports include them.

The first step to calibrating current day samples is to make sure that the sampling methods match. This means that every month, or even only every year, for several decades we will have to get ice cores and test them year after year, comparing a sample from twenty years past compared to this year and then taking a sample from the same strata a year from now and testing it. We will need to continue to resample the same near-time strata for scores of decades to ascertain whether or not there is any change in that specific strata over several decades while sitting in a dormant state trapped in ice.

Further, the charts are then often extrapolated to ridiculous proportions. When a spike is found in recent years and since the data ends at any current moment in time because there are no upper levels yet laid down, the people who make the graphs can only extrapolate the data from the most recent readings, which are all too often extrapolated to ridiculous levels. This practice is especially common in climate alarmism.

If a person is at least fifty years old and their life experience doesn't clue them in as to the cyclical rise and fall of lakes, oceans, and temperatures, etc. then they are being deliberately ignorant. I like to use the example of a child's growth. If a child grows at the same rate that they do from conception to age sixteen, then by the time a young man is thirty-two he will be twelve feet tall. Ridiculous? Yes, but that is what we have been doing in much of science when extrapolating the data.

Our life experience tells us that most people top off at around five to six feet tall. Similarly, most people that have lived long enough to experience cyclical weather changes understand that sometimes a decade is a bit warmer or cooler in general than a previous decade, and that it will likely oscillate from warmer to cooler decade to decade. Yet, regarding "scientific data", the very same people will create outlandish graphs and present them as an "accurate" representation of what is to come. It is common that these graphs end up proving inaccurate when the projected time eventually arrives.

Exaggerated extrapolation does not do well in the stock market, because the stock market will always go up and down as decades of graphs have proven. If you pay attention to "stock tips", often public stock pontificators who stand to gain from a stock spike will try to sell you on the idea of buying in, which will drive up the price and allow them to sell at a profit while you hold overpriced stock.

The point is that when you see spike in scientific graphs, you then need to ask yourself first if that spike is reasonable and if it is likely. And then ask yourself, who benefits from that spiked data? When asking those questions, you will usually find that there is some sort of convenient lie behind the exaggerated extrapolation for someone to profit due to the extreme nature of the questionable spike, as is found with the climate change predictions they keep making ten years out that when the time comes have yet to be even remotely accurate. Such lies have been occurring for many decades.

Science is Crude

Realize that our sciences are crude at best. We simply do not know what causes much of what we find with our experiments. If we did have a full grasp of these things we wouldn't be doing the experiments, which are the whole point behind science–to figure things out. But we test things and then because those things produce our anticipated results, we then believe that we understand a particular system, when in truth, we generally know little or nothing about it.

This is no different than someone believing that they understand cars because every time they get into the vehicle and turn the ignition key or push the ignition button the vehicle starts. They will arrogantly go along believing that they have it all figured out and that they understand everything about the car. But then one day they get into the car and it won't start when they turn the key. At this point they have no idea of why the expected reaction is not occurring. A person who actually understands the vehicle will be able to quickly reason through the anomaly and discover what the cause is. If the person does not understand, then they need to contact someone who does understand or they will need to spend a great deal of time researching the problem–and that is science in a nutshell.

Science is us all figuring things out and sharing those findings with the rest of the world. Our problem is that when someone finds something really interesting, as often happens, either they or we will interpret what is discovered incorrectly by drawing wrong conclusions and connecting wrong thoughts. A good example already pointed out in this book is the improper connections between "heaven" and "earth" in the first sentence, and the named "Heaven" and "Earth" mentioned on subsequent days three and four. When reading the Genesis text carefully, and while using an order-of-events-logic, the mention of "heaven" and "earth" in the first sentence are not the same "Heaven" and "Earth" mentioned on day three, logically they simply cannot be referring to the same things.

The same is similarly true of the reference to green herbs and fruit trees. The day three events are not likely the tree bearing fruit that is fully grown as many people imagine the text to indicate.

Separate the Issues

We run into additional issues in the text. For instance, consider the term "fruit", or "tree" for that matter. Translating from another language can be a bit tricky because of the nuances of the origin language and the nuances of the destination language. As mentioned earlier, Latin's "lignum" can translate as "stick" or "wood" or "tree" with each giving a very different view in context.

Just as we have been doing so far throughout this book, we need to separate the issues. Learn to separate big bang from evolution, and evolution from atheism. While these three tend to go hand in hand, they are actually all separate issues. But they generally share a central purpose of trying to explain everything as occurring without a Creator.

When it comes to believers of six-twenty-four-hour-day creation we have a similar separation of issues problem. It is important to note that *not **all*** "Creationists" believe in six-twenty-four-hour-day creation. In fact, most probably do not and will be rather ambiguous in that discussion because they are waiting for experts to finally figure it all out and offer a viable logical theory that makes sense before they make any sort of finalized conviction.

To convict on certain aspects of the first four days can be very error-prone because of our current understanding of words and our usage of the words used in Genesis One. We tend to use them in our retrospective views. In other words, a tree is a tree like we have today, and water is water like we drink today, and earth is the planet on which we walk, yet none of the language in Genesis One actually indicates that this is so. Genesis, as the name so eloquently points out, means beginning. And the Creation account is the *beginning* of much of what is established in the first four days.

What I mean is that day three only needs to be the initial germination process, which does not necessarily require seeds or fully-grown plants. Going back to the what-came-first-the chicken-or-the-egg question, cells in plants are very small items and those cells carry what we today call DNA. Using a simpler more understandable term, DNA is an *instruction* set as far as we can tell. We can take a single cell in a laboratory and "clone" it and it will grow a twin plant. We can do this by copying something that already exists. So, the "each after its kind" part is no stretch of the imagination for us.

The parts we get stuck on are the initial plants. The evolutionary process appears to answer this to the uninvolved observer. But to people who are a bit more detail-oriented, there are still missing answers; and the current view of evolution *does not* offer those answers at this time. The proponents of end-to-end evolution extrapolate things to fill the same information voids that the god-of-the-gaps rationale is often used to fill. Both methods are a way of pretending that we know what we are talking about without having to say "I simply do not know that yet."

Given that the "Seas" were gathered and "dry land appeared" on the third day, we have to assume lots of activity would have occurred that we likely would consider to be chaotic. And since water molecules were at that point likely formed as the water we call H^2O, or something very close to it, there were likely many other, of what we think of as, "chemicals" present as things settled into place and the waters were gathered making "dry land appear".

The conditions would be perfect for life to begin in such an environment.

As for the questions we didn't answer: trees likely did not grow beyond some sort of initial germination until after the fourth day, therefore there would be no growth and no substantial growth rings, but the environment would have been perfectly primed for the initial germination and growth activity.

Regarding God creating trees as adult plants fully grown, it is highly unlikely. Making that proposal is similar to saying that God made the fossils instantly, rather than them occurring through typical means. Instant trees contradict the logical actions of a discerning Creator, nor has any such thing ever been witnessed. And finally, was all of this activity on day three exclusive to our planet Earth?

It is highly unlikely that any part of the Genesis days One through Four pertained only to our planet Earth. Following the same basic principles of sequenced logic, the events of the first four days almost certainly occurred throughout *all* of the "Heavens".

Chapter 15

The Firmament

When you look into quantum physics, you will find that the deeper we peer into the elements, then the more we must theorize about what we speculate we see, or think we might see if we keep looking. What is everything actually made from? Quite honestly the answer is... *nothing.*

With a Biblical Creator viewpoint, the Creator Created and we have to try to understand this Creator and the Creation. When using skewed language from some modern Bible versions it clouds many of the issues we are discussing here, and quite honestly renders the Genesis One text of those versions useless for any purposes of scientific study. Adding the word "clouds" into the text as certain modern versions have done clouds the minds of the readers of the text—pun is intended.

We are not "cherry-picking" the Bible version that best suits our needs here; rather we are using one of the most authoritative versions available that is derived from the oldest sources available and which has not been greatly influenced in language and translation by modern science theories.

If this Creator is real and Created Creation, then we must ask: was anything there before any of the Creation events occurred? Logically we would say no, but then we get stuck in the loop of— something had to be there first, which causes us to ask how did it get there, causing us to again think something had to be there first, forcing us right back to the how-did-it-get-there question.

As mentioned in a previous chapter, "firmament" is the term that is most logical when discussing the astrophysics aspects of Creation. And if you will recall, various more recent Bible versions replace the term "firmament" and use terms like "expanse", "canopy", "vault", "sky", and "horizon" implying an extremely different picture that instantly causes new readers of the text to have an erred Earth-centric interpretation of the text. The term "expanse" is only slightly better than "canopy", "vault", "sky", or "horizon" because it takes our thinking viewpoint beyond our own atmosphere. Yet "expanse" is significantly different than "firmament" because "expanse" evokes the idea of something expanding like a big bang. Some of the many reasons that big bang is not a valid theory are further explained in *Bending The Ruler – Time Travel, The Speed of Light, Gravity, and The Big Bang.*

While there are references here to big bang, what we are really trying to determine is the order of events that likely actually occurred during Creation. This book is not specifically an argument against big bang theories. But we are making an assumption that a Creator does in fact exist, and then we are comparing the Genesis One account of this Creator's stated actions to what we actually see and understand regarding physics and astrophysics in the real world. In this way this book is more so a testament against the idea of six-twenty-four-hour-day creation than it is against big bang.

There are many views with granular details regarding astrophysics and how the planets and stars came to be, but since Genesis One doesn't get into what we refer to as gluons, quarks, and other such matter, we are going to mostly leave those finer points be. What we are working to establish is the *framework* in which those particles would have potentially been made or began.

Form Planets with Gravity

What we call "gravity" is actually little understood by us other than that the gravitational pull appears to have a force that is directly related to the quantity and composition of the periodic table of elements that make up the bodies that are being attracted to each other by said gravity. Generally, this means that an element such as gold in two large masses out in space will attract to each other with more force than would two equally sized pieces consisting of the iron element.

The various elements would have all had different levels of gravitational force and size causing them to behave differently during some of the Creation events.

When we humans don't understand things, we have a tendency to assign somewhat mystical properties to the things we do not understand. This is not a problem until we make it a sort of pseudo-religion, as often occurs in science. The patterns we see in the worlds of physics and nature are fairly consistent from the smallest to the largest particles. Meaning that, just as the electrons orbit the nucleus of an atom, so too the planets orbit the Sun, and the Solar system orbits the center of the galaxy, and the galaxy likely orbits along with other galaxies— think along the lines of fractals or repeated patterns.

And given this fractal behavior, it is reasonably fair to assume that the galaxies may also be in some sort of clusters on and even larger scale that we will never be able to scientifically detect, therefore the thought of galaxies being clustered bears no further discussion in this book because it is beyond our current scientific ability to detect.

The point is that gravity is there and is doing its thing, but what we are attempting to get at in this chapter is, when possibly did that gravity begin that would make scientific sense in the order of events that does not invalidate the Genesis One text before or after gravity's possible first occurrence? And further, it must not violate basic physics principles or basic logic. If a single term such as "sky" is used to replace the term "firmament" it invalidates a great deal, if not all, of the Genesis One Creation text. This is why this specific term "firmament" is so very important to understand because "firmament" becomes the foundation of all things.

Scientifically, we struggle with this aspect because we don't understand gravity very well, if at all. We see and feel and can measure the effects of gravity, but we still don't understand it. I am not saying that the "firmament" spoken of is specifically gravity as could be implied by the "firm" portion of the word, but rather that the "firmament" is something that is very unique and generally not typically in our normal modern comprehension of the Creation subject.

Let's look at the firmament portion of the text again:

"and god said let there be a firmament made amidst the waters and let it divide the waters from the waters and god made a firmament and divided the waters that were under the firmament from those that were above the firmament and it was so and god called the firmament heaven and the evening and morning were the second day"

Here it is important to notice that the "firmament" was named and is referred to as "Heaven" during the subsequent Creation events.

In a previous chapter, we discussed word origin. The What-came-first-the-chicken-or-the-egg type question applies not only to the seeds and the plants discussed in a previous chapter, but it also applies to the words used. If you were to describe to someone a location on Earth before us humans built our cities, you would still likely reference the name of the city and point out the fact that the location being discussed is where the current city sits today. This is true even though the city or person it was named after did not yet exist at the time beforehand that the discussion is referencing. This same effect is likely true of the Genesis text in effort to best convey the events to us. What we do not understand or know about the text is how it was given. In assuming that the Genesis One text is real and true and was given to man by God, we need to try to grasp if it was written at the hand of a man, or if somehow God wrote it? If it was written at the hand of a man then was it possibly given in a dream? Or in a vision? Or was it dictated with specific words from God to man? Therefore, we really cannot tell if the specific terms were retrospectively used until we consider a couple of events.

The first event to consider is "and god called the firmament heaven". The answer is in the naming of the "firmament" and calling it "Heaven". The Second is "and god called the dry land earth", where the answer is in the naming of the "dry land" and calling it "Earth". But now we have to wonder if the newly named "Heaven" and "Earth" are the same things as mentioned in the first sentence of Genesis One which says, "in the beginning god created heaven and earth".

We also have to consider the "waters". Interestingly, the term "waters" is always used in what we think of as the plural form and it was "divided" and later "gathered" and finally it "brought forth" things. The "waters under" and the "waters above" cannot be what we today think of as *water*. When you dig deep into ancient texts, waters can mean "fluidity" as was mentioned earlier in this book, and it is that fluidity attribute that we must keep in mind.

The view we are approaching this point from in this book is that the "heaven and earth" spoken of in the first sentence of Genesis One are some sort of pre-matter and the "waters" spoken of is a property of that pre-matter. The pre-matter was then, through some method, "divided" into the "waters under" and "waters above", at which point the "waters above" are no longer referenced specifically as "waters above". For the next point, we need only concern ourselves with the "waters under". But let us first look at "above" and "under" in Latin:

English Genesis 1:7 "and god made a firmament and divided the waters that were <u>under</u> the firmament from those that were <u>above</u> the firmament and it was so"

Latin Genesis 1:7 "Et fecit Deus firmamentum, divisitque aquas, quae erant <u>sub</u> firmamento, ab his, quae erant <u>super</u> firmamentum. Et factum est ita"

Notice in Latin the terms "sub" and "super" are used instead of "under" and "above". Is this to be understood as "super" as in ^{superscript} or is it "super" as in better than, or in some way special? And is "sub" used as in $_{subscript}$ or it is "sub" as in **sub**missive or in some way lessor than or not as superior as or as pure as?

As you can clearly see, depending upon the translator's view, we can get a very different meaning from text when translated. The terms "above" and "below" are a bit more rigid in the English language than the terms "super" and "sub" are, but can still be used in a *superior* versus *subordinate* comparison without stretching the meanings beyond any reasonable logic. Understanding these language points offers a more accurate understanding of what likely actually occurred.

In the first sentence "heaven" and "earth" were in some manner being made or were already made. They are unique and we don't specifically understand them scientifically. At this point light was made and potentially changed the earth pre-matter and possibly also changed the heaven. A distinction was then made between the light and the darkness allowing them to in some manner be separated. If light existed, we can then assume that light particles and/or light energy were possible at that time, but not yet being used in a physical visual sense.

This is where it gets interesting and things come into being:

"and god said let there be a firmament made amidst the waters and let it divide the waters from the waters and god made a firmament and divided the waters that were under the firmament from those that were above the firmament and it was so and god called the firmament heaven and the evening and morning were the second day"

Here the Creator made the firmament and divided the waters from the waters. As previously mentioned, the "waters" were likely an attribute or property of the heaven and earth pre-matter, thus allowing a separation of the waters' properties of *under* and *above*, or a *sub* and *super* type, that is to say a *sub-state* and **superior-state** type matter.

Now in an earlier chapter, it was mentioned that the "waters above" were no longer mentioned in the text of the first four days because on the third day the "waters <u>under</u> Heaven" are specifically referenced and are gathered together and dry land could then appear. This particular event is very important scientifically. I imagine that those who are deeply invested in big bang theory and the order of events as they see them might debate some of this, but it is worthy of consideration regarding the order of events regardless of any big bang opinions.

It is the gathering and the scope of view that is important to grasp here. In many modern Bible versions such as the *New International Version* that invoke terms like "sky", it greatly compromises the text. Take a look at the implications:

English NIV Genesis 1:9 "Let the water under the sky be gathered to one place, and let dry ground appear"

English Douay Genesis 1:9 "Let the waters that are under the heaven be gathered together into one place and let the dry land appear"

Remember that the firmament was *named* "Heaven"

The NIV (*New International Version*) forces us to keep viewing this from a planet Earth perspective, where we could see the waters as either rain or maybe fog being condensed and pooling into Seas. But the Douay Version has a much broader view. In stating "under the heaven" the Douay Version is much more accurate to what we see and suspect occurs throughout the entirety of space.

This gathering must **not** be viewed as large balls of mush floating in space that suddenly had water gather from within the mush to make dry land appear. But rather that day three is potentially the invocation of gravity as we know it today, which affects the elements and is caused by the elements that would have begun to precipitate due to the firmament. This likely took a great deal of time as the elements gathered in mass, eventually resulting in the heavenly bodies forming. Over time, this gravitational effect would create increasingly more fantastic events through collisions as the various bodies became evermore dense as more and more matter was drawn closer together.

We see evidence of such matter being gathered in the heavens today through our modern telescopes. In the images from those telescopes we see that there are massive clouded areas where you can see the gravity at work gathering together large portions of the clouds into what appears as massive drip-like forms pulling off of the clouded area. But let's make no mistake about it, in viewing these "drip-like" shapes, they are enormous and likely not yet as densely packed as our own planet Earth or solar system are.

Logic dictates that these drip-like formations are slowly gathering and will eventually completely break off of the clouded areas. And as they gather, the effects of the gravity of each atom will increase as the distance between the atoms draws closer and closer until the entire formation reaches equilibrium. At this point we can expect some fusion to occur on the inner atoms possibly changing the composition and also joining some atoms together creating other compounds such as H^2O.

Over time, these materials will settle into place and in a sort of sifting process will group together. You can use the visual of sand and small stones placed into a container and vibrated, thus causing the sand to sift through the small stones, But, that visual will not hold up on a massive scale. Within the confines of the forming orb that gravitationally drew itself off of the massive cloud of matter, there will be other small clusters forming.

Mathematically, we have a tendency to want to calculate these orbs as consistent throughout, and we picture the loose orb as being composed of a single-type element evenly distributed that is slowly and evenly grouping until it begins to heat up and eventually turns into some type of star and ending in an explosion casting out the various elements that were created through fusion of hydrogen and or helium, but this might not be an accurate or complete assessment of things at all.

It is likely that there are many, but not necessarily all, of the known elements in what we see in the clouded areas in space. The massive drip shaped formations that eventually draw completely to themselves into orb forms are, to be sure, very large. Their distance from our Earth is vastly further than the distance from our location to the nearest stars that are situated between Earth and these cloud orbs.

These orbs of matter are much larger than any stars that sit between the orb and us. Given the size of these orb-like bodies of elements, we can make a reasonably safe assumption that there would be much additional gravitational activity occurring within them in various regions within the orb. There could be smaller clusters of matter gathering that are larger than our Sun and the entire orb could rival our entire solar system in size by many times over.

We see the images we take from space of these magnificent structures of radiation and gaseous elements. I can come to no other conclusion that we are now witness to some of the primary steps of Creation to this very day. And based upon our viewing via our telescopes, we can be reasonably sure that it does not happen in a twenty-four-hour day.

Dark Matter

The early twenty-first century view of gravity and "dark matter" is somewhat short-sited. The early twenty-first century belief is that "gravitational lensing" is due to a gravitational pull greater than is allowed for in the amount of the detectable matter we calculate based on what we see in the region of the gravitational lensing area, thus assisting the idea of dark matter.

This mysterious undetectable matter is suspected to be the cause of the amount of gravity that is needed in some scientific calculations to cause the lensing effect. Some of the scientific figures and calculations used when trying to understand the Universe do not make mathematical sense unless there is more than what we *visibly* see through our telescopes throughout the universe.

It's astonishing regarding some scientists' apparent ignorance regarding so called "exoplanets". This is especially so when using a big bang mentality. *Everything* we observe in our solar system shouts loudly that there is a pattern of action being repeated at various levels. Meaning that the Sun has planets orbiting around it and the planets have moons orbiting around them and some of the moons have space debris orbiting around them. Additionally, there is likely other material in space that is not in orbit around stars, planets, or moons but instead is roaming space unattached to any solar system in any one galaxy, and likely many roam free between galaxies.

When referring to the apparent ignorance, what I mean is that we have absolutely no reason to project any thought whatsoever, that there are not planets around each and every star in every single galaxy, and similarly so, in all of the galaxies in the entire Universe. In fact, to assume that any star has no satellite planets orbiting it is utterly unscientific based upon *everything* we have ever seen and thus far know about science, physics, and our solar system.

When we observe a distant star, we are able, as best as we can tell, to detect an orbiting planet crossing the star's face as it orbits its star while that planet comes between our view of the star and the star itself. The fluctuation in light and the frequency of occurrence fluctuation that is observed gives us, we believe, an estimate of the speed and size of the orbiting planet even though we cannot see the planet itself. We must also understand that these "exoplanets", as they are called, that we believe we are observing, likely have moons and other debris orbiting them. All of this extra mass is missing according to some scientific calculations, which is a possible explanation of some of the mathematical discrepancies found in considerations regarding dark matter and gravitational lensing.

In addition to the mass value of the multiple planets and their moons that together likely orbit any one star, there are also potential debris belts that could easily exceed the mass of the individual solar system's planets that may orbit at a vast distance from the central star as is found in our own solar system. We would likely not be able to detect most of this other matter with any of our modern equipment. This means that the "dark matter" may actually simply be matter that we are not able to see due to its distance and the fact that it is not radiating visible light or other emission, or radiation of any sort that we are currently capable of detecting.

Then we also have to consider that there could be, for lack of a better term, "rogue" planets traveling through space that are not in orbit around, or in any way attached to, any particular star. And though they likely exist and many would likely exceed the size of our own Earth, they are mere specs of nothingness in the larger view of space.

There could be multitudes of rogue planets traveling throughout space that are completely undetectable to us because they emit no strong radiation of any sort and are just "dark matter" traveling aimlessly through space while being gently jostled about by the gravitational pull of one star or another, or the gravitational pull of one galaxy or another.

So, while the mysterious dark matter that hypothetically comprises eighty-five percent of the matter in the Universe may actually be some mysterious form of matter unknown to us humans, it could also be that it is the exoplanets and their moons along with other orbiting space debris, as well as rogue planets which is as likely, or more likely, than any other theoretical mysterious dark matter. Further, we have to consider that the idea of eighty-five percent of the matter in the Universe being "dark" is also little more than guesswork that is mostly based upon many assumptions surrounding big bang theology.

The flawed idea of big bang has so thoroughly permeated science that almost all astrophysical study is done as if the big bang occurred as an absolute event and is factual. All such theories, such as dark matter, are built upon the idea of big bang. The big bang places a distant edge limit to the Universe and assumes that if the extra dark matter mass did not exist that many galaxies would fly apart and that the Universe would still be rapidly expanding. This could be true, but again the dark matter may indeed be dark, but it might be as obvious and solid as the Earth upon which we all walk. Logically, we know that rogue planets, exo-planets, and other "dark matter" debris exists regardless of any big bang theology.

The Creation events occurring during the first four days allows for such concrete matter to occur in the form of multiple exoplanets around *every* star, likely in differing quantities along with their moons and other debris, as well as unattached rogue planets.

Somewhere between one- and one-and-a-half-million of our Earths could fit inside of our Sun. All of the planets in our solar system are only a small fraction of the mass of our Sun. The distance between our Sun and planet Earth is roughly ninety-three million miles. The star nearest to us is over two hundred thousand astronomical units away. That means that you could drop one Earth-size rogue planet about every twelve million miles apart in a single straight line between here and the nearest star and those rogue planets total mass would be approximately equal to the mass of our entire solar system.

And in rough numbers the distance from Earth to only the six-hundredth in the line of one-and-a-half-million rogue planets would be the diameter of our entire solar system, including Pluto. There could be ten times that many rogue planets and we might only have a chance of detecting any that might be nearer to our solar system if we knew enough to look, and more importantly where to look.

I am not saying that such dark matter rogue planets for sure exist, or that if they do that there are millions of them between us and the other stars. What I am pointing out is that the possibility that such rogue planets exist is very real, and the Bible's Creation account allows for that possibility by not expressly discounting it.

We have no need to make up hypothetical dark matter when real and tangible dark matter might be all around us. Dark matter need be nothing more than matter that we are *unaware* of that is not luminous. Space is very spacious. And dark matter is the big bang's gap-filling theory to keep the universe from mathematically expanding beyond big bang scientists' imaginations.

Black Holes

Black holes are another one of those areas like dark matter that need not be mysterious. We certainly have had lots of fun making sci-fi movies about time-traveling through black holes. I do not want to detract from that fun. The idea of such is interesting to think about, but this book is about real possibilities rather than fantasies.

What is a black hole? Some say it's a star that has so great a gravitational pull that it collapsed into singularity. Others believe black holes are otherwise normal, but are very high gravity stars that have a gravitational pull so powerful that not even their own light can escape their pull and they are therefore "black". Their extreme gravity will suck in any light passing nearby, either bending it or capturing it altogether. Since I am not a proponent of the theory of singularity, I tend to fall into the latter group.

A star being so massive or with such strong gravity so as to have its own light not even be able to escape it is not a stretch of the imagination at all. And further, it can be added to the list of possible dark-matter bodies that I mentioned likely exist. Considering how the lower-frequency radio waves that are a part of the electromagnetic spectrum behave where we can observe how some of the spectrum is more readily affected by Earth's gravity and environment, we have to ask if this would be any different on a massive star. Could the mass of a star affect the type or frequency of radiation able to be emitted from that star? A bit of a reach? Maybe.

What is even more interesting about a "black hole" is that black holes might not exist at all or at least some of them might not exist at all. I don't mean this in the singularity sense where the star's mass is so great that it is infinitely small and therefore is in singularity and does not exist scientifically. But rather that the black hole that we believe we are detecting simply does not exist at all.

Consider a binary star system for instance. We believe that what we see in a binary star system is two stars in orbit around each other doing a sort of celestial dance. They are both moving, yet as far as we can tell, they are orbiting around nothing more than each other. So, in the center of the galaxy where we believe we see many such stars speedily orbiting a black hole, it is possible that it is nothing more than a group of stars orbiting around each other. Just because we see orbiting occurring, does not necessarily indicate that the stars are orbiting around a celestial body in the center, though they could be.

We have an erred tendency to imagine that only planets orbit stars, and technically by those names it may be true, but that does not mean that some "planets" are not emitting light just as a star does, in which case they would be stars orbiting stars. It is very possible that there are stars with other stars orbiting them much like our solar own system, but on a much larger scale. And all or some of the "planets" would actually be smaller or lower mass stars than the main center star. Each of the orbiting stars would likely have their own normal planets orbiting them, in which case those planets would be considered moons in our perception, which is based upon our understanding and definitions which in turn is based upon our view of our own solar system.

Any two or more heavenly bodies in an orbital dance will have a different dance appearance depending upon the specific mass of each of the bodies involved in the dance. If all of the bodies are close in mass, then they will all be dancing around each other leaving a central area mostly void of the paths in the gravitational center of the group. This could then be mistaken for a "black hole".

As you can see, black holes are not likely to be stars in a point of singularity, but rather they are more likely stars with super mass so powerful that the light cannot escape them, or they are nonexistent altogether and are instead areas void central to the travel path of the stars that orbit in celestial dance around each other. Or it could be that what we call black holes are both, some could be of one nature, and another could be of the other nature, in other words "black holes" orbiting black holes in a celestial dance.

The Donut Effect

There is a gravity *donut effect theory*, implying that if a large mass is pulling in all directions then a sort of donut shape of matter could form. This is because the center will have an equal massive force pulling from middle towards the outside as the outside is pulling in. However, this is not exactly true because the forces are not quite equal. The gravity on any one side of the globe will have the gravitational force of the entire globe pulling it, which is why the bodies we see in space are spherical. Further, the internal regions are indeed being gravitationally pulled towards the outside at the same force, but those central internal regions are being pulled equally in *all* directions thus negating the ability to move outward.

However, a donut effect would temporarily occur if something repelled it internally such as an explosion, but then it would eventually collapse into a solid sphere again due to the gravitational pull. We see this effect in the images taken with our telescopes where we see large rings of matter. In the very long-term, these will likely collapse again.

Acceleration Only Appears Like Gravity

You might feel that Einstein's formulas have little place in discussions regarding Creation, but you would be wrong. Einstein equated acceleration to the pull of gravity, but is acceleration at all related to gravity or mass?

It is well known and easily proven that if you stand on a scale in an elevator during the acceleration portion of the initial movement of the elevator, that your weight will increase on the scale. You might weight one-hundred-ten pounds or one-hundred-fifty pounds, but when the elevator starts to move, your weight number will go considerably higher as you momentarily gain weight during the brief period of acceleration.

Fear not, because a spilt second later your weight will be normal again once the elevator has reached a constant speed and is no longer accelerating, but then when the elevator decelerates you will suddenly become lighter than your normal weight until the elevator comes to a complete stop.

The opposite is true on the way down the elevator. The fastest weight loss system is the initial acceleration when going down in an elevator. In this case your weight will be similarly lighter if the acceleration is similar going down as it was on the way up. In both cases the opposite is true during the elevator's deceleration period.

Does this matter? Is this a fair comparison? When Isaac Newton used an inclined plane to slow down acceleration to a measurable point he was quite right. When calculating for friction, an inclined plane can be a very accurate way of testing acceleration. Einstein, being more of a mathematician than anything, had some very good ideas that have helped many people advance their mind scientifically, but not everything that Einstein proposed is accurate.

The Einstein elevator thought experiment is interesting and we can use it to test certain data, yet it is not necessarily an accurate picture of reality. Some things can be computed mathematically, but in reality will be different, or it may be that they are misapplied.

The energy needed to increase the speed of an item may very well increase as the acceleration is increased. However, this is very different in the weightless environment of space than it is when blasting rockets off from Earth.

When in space, if the object being accelerated is free from major gravitational forces, then it will continue to travel constant at the speed at which it was moving when the thrust period ended. At the point absent of thrust, if additional equal thrust bursts are added it will speed up the object equally more with each thrust burst.

Provided that the object stays clear of massive gravitational forces, we can theoretically increase the object's velocity infinitely if we can keep applying equal bursts of thrust. This flies in the face of the Einstein theory where the mass increases as we near the speed of light, thus requiring ever more energy, with which I do *not* at all agree. This of course is with the exception of being within the strong gravitational force of a planet or star, for instance, when blasting off in a rocket while trying to leave Earth's gravitational pull in order to achieve an orbital speed as that rocket is moving through our Earth's gravitational pull.

Some calculations show that you would need infinite energy to exceed the speed of light, but with this I also very much disagree. This idea of infinite energy being required is not a proven theory and should be eliminated from classroom consideration during astrophysics lessons.

Acceleration only *mimics* gravity. Acceleration and speed *do not* increase the mass of an object in reality. This apparent increase in mass is being misread and/or misunderstood. True mass is determined by the atomic composition of the object being accelerated and it does not change in reality. Einstein, being the clever fellow that he was, is clearly in error in several of his thought experiments, either that, or we are in error in our interpretation of those thought experiments.

Just as the scale's numbers will increase as you stand on it while the elevator increases its speed, so too will the sensation of mass increase when an item is accelerated. Disregarding any amounts of energy that would be required, if we do the thought experiment in space and have a one-thousand-pound ball of solid of iron and we placed a scale between it and the spaceship that we are going to push it with, we can calculate what the scale will read based upon the rate of acceleration. It is the *rate of acceleration* that gives the false appearance of increase in mass. The mass, as stated earlier, is not actually being increased.

There is no question that when the thrust is applied that the scale's reading will increase. But when the thrust is halted the object will have reached and new greater velocity and the scale between them will once again read zero. Each time a thrust burst is applied that is equal to the previous thrust burst, then the scale will again read the same value of force. If the thrust is at a constant rate the scale will then display a constant figure, but the velocity of the object will continue to increase with each burst. This is difficult or even impossible to duplicate here on Earth because we have to overcome gravity and also overcome the friction from any substance we pass though such as our atmosphere.

In space, if our thrust force remains constant, then our scale reading will remain constant, but our velocity increases continually as long as that thrust force continues. However, in Einstein's elevator and acceleration models we are doing something entirely different. In his models, we are increasing the acceleration calculations at an ever-increasing rate, thus increasing the thrust force with the acceleration. Where in my intermittent space thrust model, the power used and the thrust put forth is identical every time.

If you want to better understand the difference, imagine my model with the rocket pushing the iron object along with the scale in between the rocket and the object. We are only exerting a minor force with each thrust, a rather gentle push. Now, when incorporating Einstein's model into my model we will have to double the amount of thrust with every thrust burst. So for example's sake, we will say doubling the thrust pressure with every intermittent thrust. In Einstein's model each thrust burst will indeed increase the velocity more than the previous but lesser thrust burst did, but that is because the thrust is so much greater with each thrust burst.

To avoid getting into the finer details that would in reality occur I will keep it simple. Let's say that our one-thousand-pound iron object is a solid iron ball and every time we apply our normal standard thrust the scale will read 100 pounds of force with each thrust. And with each thrust our velocity will increase at the same given rate of travel.

However, with Einstein's model being applied, each thrust instance has the thrust pressure doubling. The first thrust burst is 100 pounds. The second burst is 200 pounds. The third burst is 400 pounds. The fourth burst 800 pounds. And the fifth is 1600 pounds of force as we accelerate closer and closer to the speed of light.

Using Einstein's acceleration model, by the fifteenth burst we are applying over 1.5 million pounds of thrust. Now, you must realize that this is a *thought experiment* model so we are assuming thus far that the rocket and scale are some sort of super material that will not be damaged or flex no matter what, and they are pushing the solid iron ball through empty space. So we have to consider the malleability and the tensile and compressive strength of the iron ball material.

By the time we make our twenty-fifth burst the thrust is so powerful that the thrust scale would read nearly 1.7 billion pounds of thrust, yet that ball is the same iron ball of the same size and weight. And since in reality, that iron ball is iron, it is somewhat malleable, meaning that it will begin to bend or flex, in this case it will get squished more so on the scale side than on the ball's side opposite the scale. And on any thrust burst following, the solid iron ball will be physically squished further and further flat.

In this experiment we are not referring to an impact; we are ignoring that aspect in this model. On the twenty-fifth burst it would be similar to, but not exactly like, placing the one-thousand-pound iron ball in massive hydraulic press and momentarily applying 1.7 billion pounds of pressure. But for the shape of the deformation of the ball we can liken the shape to a ball of cookie dough in the oven as it melts before the baking begins. Disregarding the heat aspect, the form of the cookie dough is somewhat like the ball of iron would begin to look as the dough is heated. When the dough softens, it is more readily affected by the Earth's gravity and it begins to flatten just as the iron ball would flatten when being instantly pushed to a new and higher velocity with such extreme acceleration. And just as with the cookie dough, if the dough did not actually bake from the heat, the dough eventually would flatten and spread out quite far, as would the iron ball when given enough of Einstein's doubling of thrust pressure to achieve the speed of light.

The false impression of increase in mass is erroneous thinking. This is largely connected to the idea of relativity which carries a similar error and is that of the light's false deification that Einstein inadvertently placed upon light's assumed invincibility.

Physics is physics, and we do not need to force our blind faith of relativity or Einstein's thought experiments into it as we consider time, acceleration, velocity, or mass of any heavenly bodies, or Creation for that matter. Einstein's influence has been both a help and a hindrance to science. And the world will only grasp the hindrance aspects of those theories after we have come to realize his vast errors, or maybe our vast misinterpretation errors of his theories.

Chapter 16

Do You See what I See

Over the years I have watched many believers attempt to justify their interpretation of the Genesis One Creation account using manipulated versions of some of the theories surrounding big bang. In doing so they need to compromise a great deal and ultimately discredit themselves in the eyes of many who are deep into big bang and/or Creation research.

I like to think that the text in Genesis One is what the text in Genesis is, but as we discussed earlier, this is not true depending upon the Bible version being used. Your chosen version of the Bible will greatly influence your ability to intelligibly discuss the Creation topic. The most reliable versions of the Bible today are the Douay, the Standard King James and also the older versions from which they are derived. Using these together with the 1600's versions is a good idea. (See the book *Understanding The Bible - The Bible How-To Manual AND The Things We Don't See*.)

If you use other versions that have in them altered translations of words, then you generally reduce your ability to be accurate and you will end up grasping in attempt to explain your unjustified positions that you derived from the inaccurate text. This is when people begin to gravitate towards merging big bang and Genesis. When using the Bible versions that use the word "expanse" rather than "firmament" it is easy to get sucked into big-bang thinking because for many years it has been a publicly powerful term in describing godless creation, ending in causing many believers to ascribe big bang to God. This is because in the big bang theory, the big bang occurs in a brief instant and suddenly matter exists. For many believers this appears consistent with their interpretation of Genesis One because God did everything in six-twenty-four-hour-days in their eyes. This creates for them a good imaginary visual to ponder the supposed awesomeness of such a big bang god.

When an ill-equipped overzealous student enters college, they will put much of their energy in defending their unstable position regarding Creation regardless of which side of the debate they are on. But the believers are quite vulnerable in this regard because when in college, the evidence is stacked against them, or so it appears after their first public beat-down that they receive at the hand of their professors and fellow atheist students.

These public beat-downs are why so many students end up abandoning the religion of their youth. Some will cling to religion regardless and simply ignore that they were beat down, where others will abandon the Bible and the Creator altogether. This is why proper selection of Bible versions is critical regarding this topic and is the primary and single most important aspect in the overall Creation discussion. You will never be able to intelligibly discuss the finer points of Creation with poorly translated versions of Genesis One because they will lead your logic astray. The discrediting begins when your Creation opponents work to discredit the Bible because of the gross variations in translation found in Bibles that have been translated in recent centuries.

If your logic and thinking are led astray, then you will have a far more difficult task in *properly* connecting Genesis One to the existing *valid* scientific data that we actually find, thus causing you to grasp at straws evermore. But when you have the proper scope of view and the proper perspective view that the text was written from, you can then readily make sense of the text from a scientific standpoint. You will not need to invent things to fill in the gaps in your logic, as in "God just did it", or in frustration saying "That's what the Bible says, so that is what happened."

When you are able to adequately and properly connect the scientific data to Genesis One, then you can speak intelligibly regarding this topic. When the order of events are properly in place, then there is no need to fill gaps within the detail in Genesis One, because any gaps left can be explained scientifically without violating any of the text itself or the scientifically-based valid explanations of Genesis One or the laws of physics.

When we have the order of events and the scope of view adequately understood, then making sense of the issues surrounding light and the importance of it no longer need to be cheated in any way. When using this approach, you are able to get into the theoretical aspects of light such as was touched on at the end of the last chapter.

When an overall argument is made regarding Genesis One, and is scientifically sound, you will find yourself able to get into deeper discussions regarding light and Creation. But the earlier in the Creation text that you make an error in analysis, then the more difficulty you will encounter in any attempts to see true scientific validity in the remaining text.

Light is the first specifically stated Creation activity that we get to witness in the text. The first sentence of Genesis states: "in the beginning god created heaven and earth", implying that it was already done before the first day events occurred while not offering any additional detail on that particular issue. Though it appears to be an entirely different event, some people will include the first sentence in with the first day events, which is a fair approach, yet really has no bearing on the meaning or science of what is explained in the day one text. Nor does it specifically indicate any amount of duration between the first sentence and the finality of day one.

As discussed earlier in this book, a "day" could not possibly have been a day as we know it here on Earth because our Earth did not yet exist at that point. And even if our Earth did exist, there were no lights in the heavens to shine upon Earth to make what is our current interpretation of a day. So in the case of light being made in conjunction with "in the beginning god created heaven and earth", there is no distinction of time, but rather the order of events are being established and the meaning of the sort of "day" being referenced has no time length significance, and rather only indicates order of events. The first "day" could have been a period of an unspecified duration that could have been near to instant, or more likely it was timeless.

In considering the things that we have been discussing about Genesis One, along with the high level of certainty regarding many scientific findings surrounding astrophysical subjects, the time length issue has little or no relevance in Creation other than that the entire Creation process was *not* a short period of time and did *not* occur in an instant, although the specific commands did occur in an instant.

The initialization moment of some sort of catalyzation process may have occurred in an instant, but the entire catalyzation process likely did not occur in a moment or in a twenty-four-hour day. The first day event when light was called into **Being** could have occurred in a very short period, as could the second day events of the division of the "waters under" from the "waters above", because each of both of those first- and second-day events were likely localized for each particular particle and each event only needed each particle do their basic action in place with little or no movement from here to there. However, the obvious gravitational effects on the third day would be as we see today where they are slow and take a great deal of time to get from here to there as the elements pull together.

The order of events is different than the length of time issues are, and the order of events has an unparalleled level of importance because certain events cannot occur before other events. Some things are not possible without other things first existing. If the needed materials or circumstances caused in the early events are not available, then the later events cannot occur.

After pre-matter-heaven and pre-matter-earth were in some manner made, then light was made to **Be** and was very possibly utilizing the pre-matter heaven and earth substances allowing the distinction between the light and the darkness to be possible, thus allowing light and dark to be separated. An event such as this could be the ability that we believe we see in the state change in light from wave form to photon form and it could also be that "light" is a different realm than "dark" and what we see as light today is a mere example of that.

It is quite possible when light was initially Created that the entirety of the pre-matter "heaven" and "earth" substance would have been luminous until the division between light and darkness occurred. But in implying "luminous" it would not necessarily have been visible light as we know it today, but rather a form of energy similar to what we believe we see in our scientific experiments.

The separation of light and dark could have been an ability to cease the energy or waves and package that energy into photons or vice versa. And a part of that function could be from light's ability to travel in straight lines while not illuminating nearby items that are not hit by the light, versus light behaving like a fog rolling in and lighting up everything it touches around every corner.

Vision at the Speed of Light

Earlier I made brief mention of Einstein's deification of light, in mentioning this I am not saying that Einstein said light was God, but rather that light is given a god-like constancy as is written in his equation $e=mc^2$, or energy is equal to the constant speed of light squared which is multiplied by the mass of the matter in the equation. The *Bending The Ruler - Time Travel, The speed of Light, Gravity, and The Big Bang* goes somewhat in-depth into the speed of light and various aspects of it and why light is not constant as implied by the "c" in $e=mc^2$. Where in this book, on the other hand, we are touching on the same subject but from a Creation perspective as we look at some of the erred theories people invoke regarding light and the big bang in effort to prove their view of Genesis One as true. These sorts of rationalizations are not at all rational when you read the text of Genesis One as it was intended and as can be easily understood when reading older Bible versions that are more accurate to the original texts.

Irrational rationalizations can still be made if light is not constant, but understanding that light's speed can actually change allows far better future analyzing of facts and events.

The Light at the End of the Tunnel

One creation theory with the intention of proving six twenty-four-hour days for creation, is that when the big bang occurred it would have expanded the light or stretched it out to a vast distance in a very short period of time. The rationale behind this is that since light travels at one hundred and eighty-six thousand miles per second and we calculate a given galaxy to be billions of light years away, the light from that galaxy would take billions of Earth years to reach us. But while this theory appears to be a sufficient explanation when blending six-twenty-four-hour-day creation with big bang theology, it is not provable or even logical in real world experience. Theorizing light stretching as such, thus allowing an imaginative appearance of vast age, but actually only supposedly occurring in mere moments is a scientific cop-out. It cheats reality, actual science, and math and it is similar to saying God made the fossils sometime during the six-twenty-four-hour-days of creation in order to make the "Earth look old". Yes, some people take this path of rationalization and it cheats reality in the same way that big bang's supposed instantaneous expansion does. This dishonest rationalization must end.

We have clear, real-world evidence of light's speed when we send missions to Mars or with any other distant space mission where we communicate with the exploration vehicles. The signals sent and the signals received match what we expect regarding delay due to the given distance the signals must travel. We can very closely estimate the time a signal will take to return to us using the known light speed.

The light-speed-over-distance calculation process complicates things for the six-twenty-four-hour-day creationist's theory regarding the stretching of light. Since the speed of light is well known and is quite consistent as is explained in the example just mentioned, it indicates that the distant galaxies have been at the location we perceive them at today for a very long time, likely billions or even trillions of years.

While I do not agree with aspects of red-shift calculations and I do believe that there are many inaccuracies in many of the distance calculations surrounding our universe and its size and age, I feel quite certain that we are accurate in the scientific assumption that the galaxies are in fact very old and very far away. These distances and timeframes present a problem for six-twenty-four-hour-day creationists. Gross compensations must be made in effort to explain the possibility of Creation occurring in the span of six twenty-four-hour Earth days.

As many big bang proponents assert, the moment the big bang occurred the laws of physics did not yet exist, so the big bang would not have been subject to gravity or any other law of physics (Read the problems with this possibility in *Bending The Ruler - Time Travel, The speed of Light, Gravity, and The Big Bang*). This allows big bang to expand very suddenly at a massive rate of expansion and all in a time-slice so small that it is only mathematically calculable, but not scientifically realistic. Doing this allows big bang to mathematically occur, and it is essentially like "the-god-of-the-gaps", but instead it is done to compensate for the many shortcomings of the big bang theory. This cheating approach allows for the galaxies to fairly quickly form in their current general location giving us the time-frames we now arrive at when calculating the how far away from us a distant galaxy is through the use of light's speed.

For six-twenty-four-hour-day creationists, this is not an acceptable time frame with billions of years and is not in accordance with the interpretation of the Genesis One six-twenty-four-hour-day view of creation that they hold. To compensate for the substantial timeframes for light to travel to us, some have made a brilliant catch in the anomalies of big bang theory and assert that in the process of the initial bang the light was stretched so that it did not need to travel from here to there because it was already here and it just stretched to there. This erred and irrational theory allows creation to have occurred in any amount of time we choose to calculate.

Another theory is that God initially made the light be instantly cast across the vast distances so that it would be here and everywhere else already, in other words the pipeline or tunnel of light was mysteriously and instantly filled. This is the sort of explanation that the term god-of-the-gaps is derived from. However, given the irrational logic that is used to derive the big bang, this is an entirely fair and quite brilliant theory and it is very convincing at that, much the way Einstein's erred theories are convincing.

If you are a six-twenty-four-hour-day creationist then the light being stretched is your answer, and it is equally as valid as is the big bang. But in truth, neither theory is valid, and any error in a theory will be inherent in any sub-theory derived from it. If the big bang is wrong, then so is the six-twenty-four-hour-day stretching of light theory.

Here we have to discuss a bit more about Bible versions. Once upon a time, over two thousand years ago, there was no Bible, but rather only a collection of smaller books that where protectively carried by a group of people for many of the previous centuries and then finally compiled into a singular source.

This singular source was then carried through the centuries and meticulously transcribed from the authoritative source copies whenever it was to be duplicated. These duplicated copies where then checked and rechecked for errors. This method occurred for a very long time and has proven to be a very reliable process, as is found when comparing later transcriptions and earlier transcriptions to each other and to the oldest source copies available worldwide—The consistency in accuracy is astounding considering the potential for human error when translating so many words. (See the book *Understanding The Bible - The Bible How-To Manual AND The Things We Don't See*.)

This process changes after the fifteenth-century's invention of the printing press. Once the printing press was in place a page could be duplicated with far greater speed and with perfect accuracy, meaning that no transcription errors could occur if the printing plate was accurate. This made it possible for Bibles to be produced at much lower costs due to a far lower labor investment in each copy, thus allowing for non-clergy to actually own a copy of the Bible. Prior to that, Bibles where typically only found in churches and temples because few people could afford to have one made for their family alone.

As printing presses developed, the presses became more efficient and it became less expensive to own a press and run it relative to what they were previously able to accomplish in a given period of time. Once the offset-printing method was introduced in the late eighteen-hundreds it revolutionized the printing of any book and caused books to become very affordable to the common man. It is around this time, and just before it, where the Bible began to become common-place in the homes of many people of the general populous. This is when the issues with translation of Genesis One begin to become problematic.

With this newly found abundant access to the text of the Bible, many people began to study it and were not comfortable with the way the language flowed in the text because its dialect is not how we talk now nor have talked in the past few hundred years. Also, some people wanted to make sure that the text could be better understood so they took it upon themselves to modernize the language. These more modern versions then carry with them the translating writer's interpretation of events as they best understood it at the time of the re-write. The intentions are likely pure and might have no intent to deceive. Yet, the newly transcribed words will be influenced by the current-day culture and scientific findings at the time of the transcription and in the way that the transcribing person understood the source text.

Take a look at the publication year and proliferation of the various Bible translations. There is not much between Greek Septuagint compared to Douay and King James versions. But things change in the 1800s.

Year	Version
200BC	Septuagint
382	Latin Vulgate
1609	Douay-Rheims Bible
1611	King James Bible
- - - -	- - - - - - - - - - - - - - -
1833	Webster's Bible Translation
1860	Darby Bible Translation
1862	Young's Literal Translation
1881	English Revised Version
1901	World English Bible
1901	American Standard Version

1917	JPS Tanakh
1960	New American Standard Bible
1962	American King James Version
1977	New American Standard 1977
1995	GOD'S WORD® Translation
1996	New Living Translation
1999	Holman Christian Standard Bible
2000	King James 2000 Bible
2000	Jubilee Bible
2001	English Standard Version
2006	NET Bible
2011	International Standard Version

and more...

This obvious time consolidation between new Bible releases in the proliferation occurring during the 1800's has been greatly affected by the science of the times which has very obviously affected the language used in Genesis One in many of these translations.

The term "big bang" was introduced in the early nineteen-hundreds, but the idea was there long before. As the explosive expansion creation concept gained prominence, it was mocked by others coining the infamous term "big bang", and the name stuck. This was all occurring during the same time that these new translations were being made, and this is also when the theory of evolution was rapidly coming to prominence.

Our ConCERN-Faster Than the Speed of Light

Another physics theory area that we use to force science to agree with our interpretation of Creation is light's speed and its constancy. If you have read *Bending The Ruler - Time Travel, The speed of Light, Gravity, and The Big Bang,* you already know that light's speed is likely not constant and that the probability that something can exceed the speed of light is highly likely. And further, that light itself can possibly travel faster than the relative speed of light.

Scientists who have the opportunity to work with large particle colliders have been attempting to accelerate particles beyond the speed of light. If successfully accomplished, doing so on an industrial scale would revolutionize communication abilities for distant space travel. But this can and likely will also be extrapolated and used to rationalize six-twenty-four-hour-days creation as a viable possibility.

As mentioned in the last section, the biggest problems in the six-twenty-four-hour-days creation model is that we know the speed of light, and we know that is takes time to get from here to there. Thus, we can deduce with considerable certainty that the galaxies we see are very far away and that at the speed of light even at relatively small distances, such as from here to the nearest star, would take a very long period of time to reach us, far exceeding six twenty-four-hour days in time. This is a problem because recorded Biblical history after the Creation of mankind is only about six thousand years of history. Add to that the supposed six-twenty-four-hour-days of creation and you have the entire history of everything including creation done in six days. This does not reconcile with the mathematic calculations of vast lengths of time made when considering distance divided by light's speed. But if particles can be accelerated beyond the speed of light, then who is to say what the limit beyond the speed of light may be? This mathematically allows us to postulate any span of time we choose for the light to reach us from a distant galaxy, thus allowing us to force the calculations to agree with a six-twenty-four-hour-day creation view.

Singularity, big bang, singularity black holes, six-twenty-four-hour-day creation, and many other loose theories need to be abandoned and removed from our thoughts if we ever hope to scientifically advance as a culture.

An Eclipse

Attitude is what allows the big bang to exist in the hearts and minds of people. It is akin to those who supposedly believed the Sun to be being swallowed up when they witnessed a solar eclipse. The investigative behavior of those of us who subscribe to theories such and big bang or six-twenty-four-hour-day creation is the same as the investigative behavior that the people who persecuted Galileo displayed. Their unfounded beliefs were to be impressed upon the people, and no other possibility is allowed. Even if a theory is dead-center-on-target, if it is a *belief-only* approach it is *not* a good approach.

This sort of behavior is what caused people to do human sacrifices. "What?!" you might ask. When someone deeply and wholly believes something, they will often fight for that which they believe and will take actions accordingly. But if that belief is incorrect, then the actions they take are unfounded and may be utterly inappropriate. When inaccuracy enters into our thinking, then who is to say that the errors within the beliefs will not increase or morph into other additional even further-reaching outlandish theories and beliefs, ultimately resulting in dangerous behavior? This is how myths are made. Whether or not we want to admit it, this is occurring in modern culture socially *and* scientifically today.

Anyone who has the tenacity to actually read the entire Bible through the eyes of a non-religious perspective, which includes but is not limited to people who believe in the big bang, will quickly find much documentation from past religious leaders and philosophers, or what we now call "scientists", who contended that the heavenly bodies where just that—bodies of matter. Not gods, not magic, but rather physically substantial solid bodies of matter moving through the heavens that would obscure one another, thus such observations present incorrect perceptions to the undiscerning folks among them. This is similar to the big bang and six-twenty-four-hour-day creation theories today.

The real eclipse is the obscuring that the big bang theory and six-twenty-four-hour-day creation does to the facts as they both aspire to eclipse actual Truth.

Super Nova

Stepping somewhat outside of the needs of the Creation topics discussed in this book, since we know that light does in fact provably take time to get from here to there, we can safely conclude that a distant super nova, or other unique heavenly event that we visually see occur today, most certainly actually occurred many years ago, thousands, millions, or even billions of years ago and maybe more. This allows a Creator who would be aware of such events to cause significant human events to coincide when the light from the distant event actually finally reaches our Earth.

This allows the Creator the ability to tell us what will happen in advance regarding heavenly events. To us the coordinated human events make it seem almost magical. But in understanding this, a person could then have a reduced awe for God. However, if you are grasping what is being conveyed in this book, it should actually have the effect of *increasing* your impression of the Creator.

To be able to see an event and then tell us exactly when it will occur or to coordinate events in such a manner is quite impressive considering that the light might have occurred *thousands* or *millions* or maybe even *billions* of years beforehand.

Chapter 17

Big Bang No Matter How Small the Seed

By now you likely have a fair understanding of the anomalies in big bang theory. The most prominent is that singularity cannot occur without gravity, and gravity cannot occur without the big bang according to conventional big bang logic. Ignoring that for the time being, even if singularity could have occurred in the real world, all matter would still have existed to an infinitely small point according to big bang's singularity theory. However, if it did not exist as is inferred in much big bang talk, then it would not be there to expand during the big bang. This is big bang's most prominent flaw using faulty logic.

Some theorists say that the pre-bang matter was a small ball, maybe the size of a baseball or a basketball. But ultimately size is really a mathematically relative aspect of something. So no matter how infinitely small we can mathematically make the entire universe-ball-of-matter, we must ask exactly what is it infinitely small in relation to? The point is that no matter how small we mathematically make this ball of matter, **_it still exists_**. So, the question we must then ask is, how did it get there?

This goes back to that circular issue of when we assume something has always existed, we then wonder how it actually got there in the first place, and we deduce that it must have always been there and we then again ask how it got there.

With believers, the idea is that God always existed and that God made everything. From a religious perspective this is an acceptable beginning view. But since in the eyes of many who believe in the big bang God does not exist, so they do not have the crutch of a discerning God to fall back on.

Scientific Cop-out

Since big bang followers cannot look to a god as an initial starting point, they have to invent loopholes with which to escape their flawed ideology inherent in big bang theology.

Big bang is in essence a scientific cop-out for not having to accept the concept of a scientific God, but also for not having to explain the anomalies in everything believed to be true in the big bang theories. The idea of an expanding universe appears to make sense and is easy to picture in our minds without much effort. This is because most of us are familiar with explosions and how a tiny piece of material can make such a big ruckus over such a large area in such a short period of time.

As a matter of fact, some of the earliest expansion or explosive creation theories were not necessarily intended to be godless. There have always been people who were wondering regarding specifically how it all occurred. The idea of expansion is one of those early thoughts of how God might have done things during the Creation period.

But for all of the reasons mentioned throughout this book, a big bang is highly unlikely to have caused any of Creation. Just as many six-twenty-four-hour-day creation believers have their "gaps" as a cop-out, so too do those who adhere to big bang theories. The idea of cyclical bang expansion and contraction and expansion and contraction endlessly repeating is one of those cop-outs. This is because we're still left with the question of, **how did it all get there to begin with?**

Darwinian evolution and the big bang are cop-outs for people to believe in when they can't find their way clear to explaining what they have heard or read or what they see. Similarly so, the Bible is abused or believed in the same way. But this doesn't mean that what people believe about what they have seen or read is actually true. For instance, just because we believe we see "red-shift" does not mean that the universe is expanding, and just because the Bible says that light was created on the "first day" doesn't mean that it was a literal twenty-four-hour day. As a matter of fact, if we assume the text is accurate and make the assumption that we are likely interpreting it incorrectly, then we can more easily realize, scientifically, that the text talks about a "day" before a day as we understand it today could have existed.

Many "scientists" claim that the Bible is fables, fairytales, or myths, yet many have never read it completely cover to cover. Claiming the Bible to be fables or fairytales is as absurd as believing that the Sun goes around the Earth. Many Scientists who have actually read the Biblical text have only done so with disbelief and prejudice in their heart and mind while misinterpreting it all along the way.

It is often said that you need to have special abilities to understand the Bible, and only certain people can interpret it. And to an extent I believe that this is true, however, those "certain people" can be any of us provided that we are actually searching for the truth rather than searching for what *we* want to hear.

Whenever we create a cyclical trap to help us mentally cope with the questions of what was there before creation, and if it was nothing then how did everything get here, we trap ourselves with our rationale in that endless cycle. Big-bang-oscillation, singularity, and non-applicable laws of physics are all scientific cop-outs that far exceed any god-of-the-gaps mentality.

What is Negligible?

Calculating big bang down to negligible levels as is done with singularity by saying that the laws of physics *briefly* did not exist at the very first moments of the bang using a time slice so small that it is not even worth considering; doing so places both ideas in a category of "negligible". Just to understand how small a time slice big bang believers propose those initial moments were, consider that it is speculated that the expansion was trillion-trillion-fold in under a trillionth of a trillionth of a second, or so I have read. When we can negate our gaps in our minds to so small of a point as to be negligible then we no longer have to deal with explaining the anomalies that exist in a theory without the full and complete negation of those two thoughts. This scientific cop-out clears our minds and allows our math to work where it would not otherwise function as desired—it is cheating on a grand scale.

Then we have another negligible big bang aspect which is the Wilkinson Microwave Anisotropy Probe (WMAP) energy map. In theory, the WMAP takes temperature readings of areas of space that carry residual heat energy left over from the big bang—in theory. The high and low temperature variations between the colored regions on the WMAP are said to be about 0.0002 degrees kelvin or 0.00036 degrees on your house thermometer. That is about four one thousandths of a degree—a negligible difference. However, this difference does appear as if it is actually being detected. So, does this temperature variation matter? And what is the overlay repeatability WMAP imagery data?

Regarding the repeatability of the WMAP, I was not able to confirm if it was repeated or repeatable; or if it was repeated, then whether or not the maps perfectly matched in an overlay scenario. If two individual series of scan events had been done and it was overlaid with an accuracy substantially exceeding sixty or seventy percent, then such scans hold some legitimacy. However, if an overlay is fifty percent or less then it can be considered random regarding the placement of the regions with higher or lower temperature readings and should then be mostly disregarded as junk-science. Though when we invest several hundred million dollars into such projects we had better find something that appears legitimate, otherwise funding for further similar missions will be much more difficult to come by in future grant requests.

Regardless of whether or not a second scan was done to check for accuracy and then overlaid for accuracy comparison is certainly one point of scientific honesty to question as to whether or not that occurred, but more importantly, the actual meaning of the data is still a question even if the temperature variations have a reasonably matching overlay.

What Does the Data Mean?

There are several issues with the WMAP readings. The first is that the readings are two-dimensional. While the scanning satellite scanned the entire surrounding view of our Universe from its particular location during the scan period, the scanned temperature cannot determine depth of scan, so we have to assume that it was sampling somewhat throughout each scan-area-pixel or sensor pixel line-of-site and averaging what it sensed. Whether or not this would be intended or is simply the result of the chosen devices doing the scanning, it is nonetheless what happened. If we send out signals to bounce off something then the negligible 0.00036 degrees Fahrenheit WMAP reading could not be considered reliable due to the potential noise from the returning signals. On the other hand, if the sensors are receiving the energy coming towards it from space, then the readings are irrelevant because they are not in a static position and are in motion and are received as radiation coming to the WMAP sensors.

The point of the WMAP was that if a big bang occurred and expanded, the massive amount heat and energy created at ignition would be spread out, and when stretched out would cool due to the expansion. When a given amount of energy in a single location is stretched it will proportionately reduce the energy in any given area of the entire size of the outstretched energy. Think of this in terms of holding your hand near a light bulb and the further you move your hand away from the bulb the heat will be reduced because it spreads out and stretches the further from the light you are. To better understand this, we can reverse it, by picturing the sunshine going through a magnifying glass where when the magnifying glass focuses the sunlight properly it will begin to burn a hole in a piece of paper. What is occurring is the heat that you feel on your skin on a sunny day is also shining on the magnifying glass and the magnifying glass takes all of that lower temperature heat that is hitting the magnifying glass and concentrates it into a very tight spot on the paper, making the spot very hot—hundreds of degrees, so hot that it can burn something. The reverse of this is what occurs when you stretch energy out again, and that is the theory behind the analysis of the data collected that is believed to be residual big bang energy detected using the WMAP device.

As we speculate on creation methods, such as big bang expansion, we make certain assumptions of what we expect we might find when looking for certain phenomena. The WMAP is believed to be the confirmation of one of those assumptions, but that ***does not*** make it a correct assumption.

Scientific Proselytizing

We spend billions of dollars on our assumptions from which we derive our theories, and when that much money and people's credibility is on the line, we are going to make it work, even if those who benefit from it all have to force it or even lie about it in order to do so. With the big bang, the idea is that if it occurred then there should be residual energy left and we should be able to detect it. We are told that it is this residual big bang energy that we see in the WMAP. This 0.00036 degree Fahrenheit temperature difference (2778 x .00036 = 1 degree Fahrenheit) between regions may actually exist as the scan representation shows and we are often told that this finite variation between the cooler areas and the warmer areas "proves" that a big bang occurred. This sort of scientific proselytizing is used to cover any embarrassment from the highly questionable readings and/or conclusions, and it is a swindling of the people rivaled by nothing before it.

The WMAP is far more likely to be showing variations of the temperature that the light energy causes as it travels through space to the sensors than it is to be showing residual energy from a big bang that almost certainly did not ever occur. Further, if the energy is actually there as stated, then how do we know that it is from a big bang? *Any* creation method is going to produce energy, and if that energy is unable to fully dissipate then it will be detected with any instrumentation that is sensitive enough to pick up such insignificant temperature differences. And any thought that it would be a perfect evenly distributed energy if the Universe was static is a childish belief of monumental proportions.

All too often, Biblical-science proselytizers take correct scientific findings and adopt them as if those findings are their own. They then go on to force those findings to align with their desires. But to do so is usually contrary to reality, because real science will be homogenous with all reliable information and does not have to cheat or force anything. Proper interpretation of a true and proper Bible must match *true* scientific findings, but to do so we have to stop believing scientific lies and inaccuracies.

What we find, and what we concluded about what we believe we have found, are often two very different things.

It is Difficult to Convey a Message

If the six twenty-four-hour day creationists continue to adopt big bang principles and continue writing those principles into their Bibles through further erred translations of the text, then they will be blamed for the error-filled big bang theory when it is finally realized as the false doctrine that it is and that it fails *all* legitimate scientific-calculations. Then to appear innocent, non-believers will point the finger of blame at Christian's for perpetrating this hoax onto the world, a hoax perpetuated by a priest. Beware big-bang-Christians, because they will be coming for *you* soon!

First, some Christians reworded the Bible with foolishly inaccurate words in some of the more recent Bible translations. These inaccurate translations include the influence of the erred science of the time in which the translations took place. Then many who bought these poorly worded works of folly read the text and were able to begin to more readily mesh the modified and compromised Genesis One text together with big bang theory, thus imagining that they were using a "scientific approach" to understand Creation. Doing this gives the erred impression of legitimacy, but it is not legitimate–Placing error upon error is *not* the way to Truth.

It is difficult to convey a message if someone cannot comprehend the fundamental aspects of a theory. That's why big bang is so attractive to so many people. They can picture it in their minds, but they are not forced to think through the difficult parts that would otherwise occur. For some Christians, trying to tie the big bang and six-twenty-four-hour-day creation together removes some of the gaps that the god-of-the-gaps accusations include.

As we consider the various ideas about Creation and look at the Genesis Creation text, and then realize how limited it is in detail and how broad the statements are, we must realize how difficult it is to convey large volumes of complex information.

This task is difficult enough when trying to convey such information to trained people who already study astrophysics science subjects. Imagine how thick the Bible would be if the finer details of Creation where all written into the Bible's Creation text. Then consider all of the concepts of the functionality of the things we cannot see such as electrons, neutrons, protons, and photons and the sub-particles that compose the particles just listed. Now add to that the complexity in translating it all. We can't properly translate the very limited Genesis One Creation text that we currently have as it is. Do we imagine that if we can't even get a term like "firmament" correct that we would somehow get hundreds, or even many thousands, of pages of far more complex text correct, text that would be describing things we would have had no words for?

The Creator has given us concise message after concise message which are all written in the Bible, and we can't even follow those simple instructions that were given for our very own protection. Think of all the imaginary things people might make up when trying to interpret a Bible filled with complex scientific thoughts.

We have all of these wonderful symbols called letters that are used for writing, yet we have a hard time accurately conveying our own thoughts with them. Now imagine trying to do it all without letters. The Creator has no symbols to automatically give to you other than your own mind, and your own body which is a representation of Creation.

It is even more difficult to convey a message when other false doctrine resides in place of truth in our heads and in our hearts. Stay clear of becoming a false preacher of creation. Strive to be *accurate* rather than *"right"*, and then you will be on your way to beginning to understand actual Creation.

Chapter 18

Let There Be

Who is this "Creator" that Created the seen and the unseen merely by speaking it into being? Is this Creator some white-haired elder sitting on a throne in Heaven with a magical box of clay with which to make all things? Does this Creator even exist?

What we all need to understand is that this Creator to whom we have a tendency to attribute much magical hocus-pocus, is not like us in a physical sense. There are no physical hands with which to do the work. If there was actually no big bang or any other self-initiation of creation, then Creation was Created and it was done without hands, which is critical point to consider.

The Way It Came to Be

Words are a very interesting thing. What are words? What do they mean? Where did they come from?

Words ultimately have specific underlying meanings, but because there seems to be no limit to situations and items and events, etc., words have a tendency to continue to obtain alternate meanings in our eyes and are used to describe other thoughts that are similar to the original thought of the word. Such similes are common and without them little communication would occur.

Words are built with letters, and letters are symbols, and each of those symbols has a phonetic value or sound. Then each symbol, when combined with other symbols, can have a somewhat varied new sound. Additionally, these sounds can also be compounded. When several of these sounds are put together they become a single word. Each word has an underlying meaning and these meanings will be used to share or convey what we call "information". This conveyance of information is what we commonly refer to as "talking".

Take a moment to think about what was just said and how all of this really works. If you ponder it long enough you should begin to be astonished by its efficiency.

This process of relaying information and thoughts from one person to another person is really one of the most incredible inventions ever. We can debate about evolution and big bang versus Creation, but how do we actually accomplish that task? Meaning, how do we communicate to be able to even debate about such questions? The complexities of communication are vast and we are not really anywhere near figuring it out as a world of people.

At the end of the twentieth century when computers were becoming more and more able to speedily do their calculations reliably, artificial intelligence was on the horizon. Without the ability to process data quickly, a computer has little chance of at least appearing intelligent. But with processing speed comes the ability to plow through massive amounts of information in a very short period of time.

When a reasonable appearance of artificial intelligence was achieved, the computers were able to use certain types of logic that were pre-programmed into them and through that logic were able to offer responses that are reasonably accurate and very useful to man. As this technology increased, it became commonplace that such devices would be available to us to assist us in many day-to-day tasks. This convincing technology must be understood as *artificial* intelligence and not *true* intelligence.

However, when discussing *intelligence,* we come to a point where we blur the lines due to the convincing nature of the technology. Everything regarding intelligence and automation that we create is merely an attempt to copy ourselves to assist in our work and in our lives in general. For those who do not understand or work with computer programming, artificial intelligence is considerably more convincing. When you are involved with and understand the various technologies producing such artificial intelligence, it takes away from the perception of the humanlike nature of the technologies. This is similar to the magician example mentioned some chapters back.

Understanding the technology is much the same as understanding how a magician makes an item disappear. Once you know the trick, it loses its ability to amaze us in the same way it tended to before the trick was understood by us. But let us not misunderstand this effect because the magician is still a master at the craft, as is artificial intelligence really quite impressive and very useful to us. Understanding *how* something occurred should not take away from our awe of that technology. In fact, seeing and knowing how it all works should, if we are honest, impress us all the more.

I have witnessed this sort of fall from glory when someone is amazed when they see something, but once they begin to understand how that thing occurred they then adopt an attitude of arrogance and then disregard the master's work. It is this same behavior that we see in the sciences where when people begin to understand the basics of the inner workings of the Created thing being studied, then with some people there is a tendency for them to be inclined to move towards the belief that there is no God. Knowing how something occurred does not mean that you can do it too, which is the greater point to be taken here.

At some point, each person has to grapple with the "Is there a God?" question. And each person has to eventually make some sort of decision regarding that question. Now, it is important to point out, and this cannot be stressed enough, no matter what you or I choose to believe, it will not change the truth of the actual true answer to that question. If this "God" *does not* exist then nothing we say or do or believe will change that fact. But the opposite is true as well, if God *does* exist then nothing we say or do or believe will change that fact.

Since this book, *The Science of God Volume 1*, is about checking the viability of an astrophysical scientific Creation through the intentions of God, we have to consider that the Created matter was caused to exist by a Creative discerning Existence.

Is Everything is Made from Something?

Scientifically speaking, we assume that everything is made from something, but this is not true. It cannot be true unless you have chosen the slippery slope of oscillating-imagination. "Oscillating-imagination" refers to our human inability to imagine that which in our minds cannot be. This is what was mentioned earlier regarding the what-came-first-the-chicken-or-the-egg dilemma. If the Chicken came first, then how did that happen, because eggs are where chickens come from. Okay, if the egg came first then what chicken laid that egg? We face the identical oscillating dilemma with Creation when trying to rationalize it without placing a Creator in the equation. The questions of Creation are not removed by saying "The Creator did it." because we are now, and always have been searching for *how* it was done, such as with a big bang, or by saying "let there be"—or by any actual plausible means.

In our human nature, if we can trivialize any aspect of Creation, then we are more willing to overlook that trivialization in order to rationalize our desired belief. In science we do this by digging ever deeper, hoping to find some base substance that we could imagine has always existed along with the explanation of how it always existed. And as the curious group that we humans are, we have been doing a fantastic job finding, or at least theorizing, ever smaller component parts that make all things. But one pesky problem keeps reoccurring and dashing our hopes of having science be entirely godless, and it is the-chicken-versus-egg-problem. What was first, but more importantly, *how did it get there?*

Our human minds have this unique ability to wonder in a way that is just different and far superior to that of any of the animals. And this unique ability always comes back around to the fact that no matter what the substance is, or how small it is, or how seemingly insignificant it is, our minds know that it could not always have been there. Everything that we know or have ever experienced or see is always causal. We have no reason whatsoever to imagine that this is not so for that which came first—it came from a cause.

This is why the idea of a Creator is so appealing to many six-twenty-four-hour day creationists. Because if the Creator always existed, and if we accept that fully, then the initial production of the most based aspect of matter was Created by the Creator and thus there is no more need to discuss it any further.

With those who believe in the big bang, their method to answer, or rather ignore, this question is the theory of singularity. If we make everything so small that it is in essence non-existent, then we have solved the where-did-it-come-from issue. However, as mentioned earlier in this book, no matter how small we mathematically make the singularity, and no matter how many zeros lead the size after the decimal point, it will still mathematically exist. So how did it get there?

If we cut the nonsensical ideology and approach this like adults, then it will be understood that the idea of big bang cannot ever answer the question of what came first because it is an utterly impossible theory and it should be abandoned. This is why human rationale generally tends to lean towards a Creator. When the dust from senseless debate about impossible theories clears and we come to realize that something had to **cause**, then and only then can we begin to answer this most real of questions—What was first?

At this point we are going accept that there is a discerning Creator who is not subject to physics as we understand physics, because all signs of Creation point to the fact that this Creator Created Creation and it is this Creation that the laws of physics are derived from. Let us understand this as—it is Creation that dictates the mathematical formulas that we have invented to describe Creation—it is **not** the other way around. If gravity worked differently than it does today regarding its force or pull, then the formulas that we work with would be different when calculating those gravitational forces. Understanding this should clear your thoughts considerably when pondering Creation topics.

Upon accepting the Creator as an actual entity, we are still left with the same nagging question about Creation. How did it get there? God did it, right? Yes, but how? And it is the "how" that is now, and has always been, at the core of this topic.

We humans have a unique ability to somehow understand that it all has to have come from nothing. And since the Creator is not substance, we still have that question of, "*How* did it happen?" What answer would be satisfactory to you that would make sense to a point where you were satisfied and no longer had to oscillate your thinking—the chicken, no the egg; no, the chicken; no, the egg...

Not only is accepting the idea of a Creator the easiest way out of this problem, but it is also the single most logical perspective that has ever existed. This will be elaborated on in a later chapter, but God did it with mind. But, what is a "mind"?

Existence is Not Subject to Time.

For you and I, does time before us or after us really matter? Since we did not exist and will eventually die, then only the time we are here matters in a testable scientific way. If the Universe is billions of years old, or if it is trillions of years old, does it matter to us? Not really. In fact, if there is no Creator then nothing really matters at all, but if a discerning Creator *does* exist then there is the issue that stands outside of physical science which is the idea of an afterlife.

A Creator that Created Creation would have been there before the counting of time and logically would not have had any sort of watch to gaze upon in order to know how long something took to occur. In fact, it is entirely possible that there will not ever be a way to know the age of the Creator because the Creator possibly does not know and cannot equate that length of time to anything.

If *you* were floating in space and had no time-keeping devices and you could not feel your heartbeat and did not need to breathe with regular breaths, then you would not have anything other than the stars with which to gauge your time in space. Now imagine no stars, no movement, no planets, then how would you gauge length of time?

The entire Universe is a gigantic clock with invisible pinions and wheels that we call axis and orbits. These move in slow patient synchronization, and whether or not it all existed, the Creator still would exist.

I touch on the subject of existence in other books, but ask yourself this: What is existence? Can existence be measured scientifically? The answer is yes and no, because it really depends upon what we are looking at. When using science, we are always looking at the physical realm. But with our minds, things get a bit more **in**tangible. This is where the afterlife enters the discussion, but we won't get too deeply into that in this book.

What is a mind? What is thought? What is the Creator? Some people believe that when our bodies are done, then our mind will possibly somehow continue in the form of what we typically call a "soul". Many people accept this, but there are those who forcefully reject that souls and an afterlife exist. When we speak of Creation, it is that "soul" or "mind" that is made in the image of the Creator. So, what is it? What is the "mind"? We can detect electromagnetic activity from our brains, but that is only produced because of our thoughts. *Thought* is **in**tangible.

Based upon the patterns we see in nature, if we do in fact have a *soul* or a *mind* which will live on after our bodily death, we can assume that that our mind is made in the pattern of the Creator's mind because we are "created in the image of". And it is this Creator's mind or ability to think that has produced the "earth" pre-matter described early in this book. This Creator had eternal past in which to invent and Create what we see today. From a Biblical perspective, we are told that we have similar ability, yet too often we mock that as myth or fantasy, and yet there are multitudes of people that have utilized this ability to alter their own body physics.

Existence is not subject to time. You either exist or you do not exist. Having existed an hour ago is not the same as existing right now, and existing in the future is not the same as existing *right now* for any new moment in time. In our concrete tangible world, we tend to measure existence with time, but existence is better measured in terms of truth or no truth. It is either there or it is not. The soul cannot be measured scientifically, and if we do in fact have souls, then those same souls are in the image of the Creator who cannot be measured scientifically and is timeless because of the fact that the Creator is a non-tangible and is thus immeasurable Spirit.

The eternal past was potentially available for the Creator to conceive things to be at a higher level, thus allowing a way for that mind or spirit to develop and subsequently draw pre-matter-earth together in a way that divided it and eventually created locations of particles that **_became_**. This commitment is why it all is here today, and if you accept this basic premise, it can lead you to understand a great deal more when you allow it to do so.

Singularity Cannot Exist Without the Laws of Physics

Singularity was mentioned several times earlier, but let's elaborate a bit more here. It is a sad case when people work to attempt to tie science and Creation together through the use of inaccuracy and lies via the dishonest big bang theory.

There are too many problems with big bang to mention in this book and some more are touched on in *Bending The Ruler – Time Travel, The Speed of Light, Gravity, and The Big Bang*. But the single most prominent issue with big bang is the eyes-closed attribute that the "laws of physics did not yet exist". I do not disagree with that. In fact, it would be utterly ignorant to assume that the laws of physics existed before Creation because the "laws" are, in fact, the *result* of Creation rather than the *cause*.

Once the base components of matter were laid down, then the laws of physics resulted from the laying down of those base components. This is fully logical and few people would be foolish enough to argue such a sound logical approach.

But it is a problem when we imagine that these laws of physics did not exist before the big bang if we are to accept the big bang as a plausible theory. Setting aside the fact that the big bang is a horribly flawed theory, we have to consider the simple logic that even young children can see. If singularity requires gravity calculations in order to mathematically arrive at singularity, then we cannot make such a conclusion because the laws of physics did not yet exist according to the theory. And therefore, singularity cannot exist no matter how small we make the time slice of the supposed non-existence of the laws of physics.

While similar to the chicken versus the egg question, which is a what-came-first type question, the non-existent laws of physics problem is easily solved because it is with far-reaching mathematics, and with far-reaching mathematics only, that the big bang singularity can even ever be achieved. This achievement cannot occur without the very laws used in the calculation thereof. If those laws ceased, then who is to say when they ceased? How big was everything when the laws were able to begin?

As you can easily see using even the slightest bit of logic, singularity cannot exist without the laws of physics. And therefore, the big bang is clearly a big dishonest theory.

Chapter 19

An Open Mind

Open minds make for great science. Closed minds are ignorant and will always endeavor into folly. We can confidently disregard big bang because of nothing more than the issue regarding the theorized non-existence of the laws of physics used to derive the theory.

However, the non-existence of big bang and to dash away that ridiculous theory does not in any way prove the existence of a discerning Creator. So then is there any way to "prove" that a Creator does or does not exist? That is all going to depend upon how you use your logic, whether you're going to allow your thinking to oscillate between chicken-versus-egg, or if you have an open mind. Short of having some sort of God-detector, there is no way by testing with instrumentation to prove the existence of a Creator. But certain evidences do exist if you are willing to attribute them to the Creator and the existence thereof.

A Parable

A scientist leaves his house to study the large forest that he lives in, when an acorn hit him on his head as he walked out the door. Now he wonders how this could happen. Because of this peculiar event he decides to dedicate his life to the study of the source of this acorn. After many years he knew that he had discovered that he had been living in an oak forest. His firm conclusions where published for the entire world to read. They described how the tree was the source of the acorn that hit him on the head and how it had grown and cast its seed upon the ground where only the strongest trees would survive and grow.

He theorized that this acorn was either made by God or it somehow spontaneously grew into an acorn and then into a tree. As the acorn expanded into a tree, other acorns began to precipitate out of the tree. From those newly formed acorns more oak trees expanded causing more acorns to precipitate causing a chain reaction that produced the entire oak forest that he had theorized he lived in.

It was settled, he knew for sure how the forest was formed and that it was an oak forest. His papers were read and adored by many young people who were also interested in how oak forests came to be.

Decades later a bold young scientist decided to further study this oak forest and she began to notice a difference in many of the trees. Most of the trees did not cast acorns upon the ground but had varying types of seeds. She began to speak of this with colleagues but was ridiculed terribly for her audacious proposals. All of her peers were firm believers in the fact that the entire forest was filled with oak trees. After all, the now respected scientist who began the original study had presented many acorns as evidence that proved that it was scientific fact that it was indeed an oak forest.

As the years drew on, new scientists began to wonder at this forest and eventually built a rocket to use to study the oak forest. They were certain that this would once and for all discredit the scientist who made the audacious claim that it was not an oak forest. When the rocket had returned its pictures from above, the young scientists had confirmed that it was indeed an oak forest. The trees where all a similar shade of green and were dispersed relatively consistent throughout the entire forest. They were able to estimate the age of the forest and the size of the forest. It was settled, the distant upper mission rocket proved what they had already known–it was an oak forest!

Even though the initial scientist could not see the forest through the trees because of his intense study of the tree from which the original acorn fell, those who followed him did see the entire forest by use of their distant upper mission rocket and they believed that they proved beyond any doubt that the forest indeed consisted of all oak trees. It was solved and considered scientific fact and taught to generations of students.

But it was not an oak forest; it was a maple forest with a small oak grove at the edge of the forest that surrounded the house of the initial scientist. The forest's age was misread because the land around the forest was planted with wheat every year thus not allowing the forest to continue to expand, thus causing them to underestimate the age of the forest. The bold young scientist was eventually begrudgingly credited with the discovery that it was not an oak forest. Yet some still refused to believe this to be true and continued setting out to prove that it was an oak forest. Eventually they decided to walk through the forest and study the ground and found many acorns lying throughout the forest. In fact, some areas had piles of acorns; therefore, they knew that it had to be an oak forest–But it was squirrels that carried the acorns through the forest and piled them up–Our theories are often *not* correct and the squirrels will eventually gather the nuts.

Your Scientific Mind

We are all going to believe what we believe when our minds are closed to truth. You cannot claim to have a scientific mind if you disregard other plausible possibilities. This does not mean that we must suddenly change directions. We need to pause and step away from our chosen theories for a moment and consider other theories and then test and compare them to see which theory is the most logical, if any are at all.

Atheists often claim that they are using the "scientific method", when in truth they are not. Christians have ignorantly allowed atheists to define the concept of "scientific method" to exclude anything that the atheists cannot reconcile with their atheist beliefs. To this thought an atheist scientist will typically become agitated and go on a rant of how the scientific method uses only "evidence". To which I ask, is the big bang based upon either of the scientific method or evidence?

Six-twenty-four-hour-day creationists will similarly claim that they are using the scientific method until something cannot be explained, and then it is put in the God-did-it category to fill in the gaps.

Neither of these two perspectives is particularly useful, and neither truly uses the scientific method. The scientific method consists of much examination and reexamination and then removing aspects that are easily disproven or are utterly illogical.

If you fall into one of these two categories where either you have decided that there is no God, or you have decided that God did it all in six twenty-four-hour days, then you simply do not have a scientific approach or a scientific mind. A scientific mind requires consideration of alternate ideas. You do not have to see it their way, but if you do not take other theories into serious consideration, then you will not ever see the truth when you are wrong because your foolish mind is already made up based upon your blind faith.

Even if you are correct in your chosen theory, you still must consider alternate theories because your current theory could be wrong. Some people would consider this doubting, and I suppose depending upon your attitude it could be doubt, but double-checking your theory through comparison is a wise course of action, lest you get caught up in some other instant creation myth akin to big bang. Double checking also shows confidence and courage in your own theory when you actually allow your theory to be challenged.

Both the six twenty-four-hour days creation and the big bang share one common aspect—they both occurred quickly. Yet, if you really take a look at the evidence, quickly is perhaps the most unlikely attribute of Creation, especially the early stages of Creation.

If we are Created in the image of the Creator, then we are capable of far greater understanding than what we scientifically have been displaying over the past few centuries. The Creator would have to find a way to Create from nothing, which likely took enormous amounts of what we call "time". The mind of the Creator is likely indescribably old and would have had an eternal past to find a means of thought with which to Create with from nothing, and eventually did so.

The Way We Learn

There is a difference between the way many of us learn. Sure, it's always through experiences, but the way we each internalize those experiences differs somewhat from person to person. Our experiences include our learning at schools and such, as well as our experimentation throughout our entire life.

Some of us heed instruction and are able to repeat with a high amount of accuracy what we are taught, where others will tend to look more at function.

Those who heed instruction will need instruction always and can accomplish much when adequately trained.

But those who look at function can understand how something works and will be able to accurately elaborate on it; and if possible, improve upon it. They can more readily predict what will occur without having to first test a hypothesis. They are able to predict because they have an ability to quickly examine the mechanisms that are in place and project the cause-and-effect aspects of each part in mere moments. Those who understand function are typically those who invent and create the things we use every day.

To understand this better, some people will memorize the multiplication tables as children and it is more through memory that they can multiply quickly, where other people will multiply through function. Multiplying through function takes a bit longer as a child because they then understand multiplication in a rows-and-columns manner and picture it so and then will calculate based upon that mental image. Memorizing, however, eliminates that but becomes more difficult when the calculations grow in size. So, typically, someone who understands function can estimate more quickly and often more accurately than someone who only memorizes their math tables. All learning works similarly.

Methodic Learning

We learn and discover incrementally, and then eventually the false things are clearly and easily proven wrong through function. But if we never learn to learn through function, then we can be easily deceived. Sometimes, even theories that are true are believed in a wrong manner, and when those theories are challenged because there is no basis for the belief, those accurate beliefs are then subsequently abandoned by the believer due to that lack of basis. But if you learn using an approach of *function* rather than through *instruction* you can easily discern errors in ideas that oppose truth.

Let's take for instance the six-twenty-four-hour-days creation approach. If everything was created in the brief time period of one hundred and forty-four *total current-day hours*, as some people attempt to describe the creation events, we are forced to invoke the God's-hocus-pocus-like-poof-it-is-made! perspective. Now we have to ask ourselves if we really believe that this all-powerful Creator is going to actually want us to believe *"just because"*, or might the Creator want us to at least try to actually *understand?*

Do we really want to imagine that an all-powerful Creator would hold the position that we have to believe for the sake of blindly believing? That is highly unlikely.

When we learn, there are specific methods that can be taught. These methods have been sought after for a very long time, but the problem is that if you have not been taught to learn through function, then you cannot teach people to learn through function. In fact, if you have not been taught to learn through function then you likely do not realize that the "function" learning method exists. A problem with most schools is that they do not have enough teachers who understand how to teach children to learn through function; therefore, many students suffer at the hand of teachers with inadequate skills. Most teachers, and therefore most schools, teach through instruction only, thus causing the students to always need to be instructed and told what to do, and eventually, out of frustration, many begin to behave badly at some point in life. This problem will become worse with each subsequent generation if the memory method of learning is not guided at least to some extent towards learning through function.

Keep your mind open and begin to understand *function* and pass that skill on to your children. Doing so can improve most, if not all, areas of a person's life.

Having an Accurate Perspective

Our assumptions about our own perspectives of what we see about Earth and what we see in the heavens/cosmos and how we view the Bible and scientific evidence with severe prejudices is our human downfall. We humans gravitate towards a personal perspective, always seeing things only through *our own* eyes. When we learn through function we are better able to view from any point within a sequence of events and get a reasonably clear view of how that part of the sequence occurred, and thus what will likely occur next.

Humanity's biggest problem in trying to put together the puzzle pieces of science in harmony with the Bible's Creation account is our *perspective*. Too many of us on both sides of the discussion believe that there is only one perspective, which is that the Bible and science do not belong together.

But I propose that *the Bible and science* **cannot be separated**. To get an accurate perspective about science and the Bible, you must view the Biblical Genesis Creation text as being described from the Creation or Creator viewpoint. I assure you that it is a very different view. If you have ever flown in a jet at five miles up it will give you a good idea of how differently the view is between a personal view from the ground compared to a view from 35,000 feet in the air. How much more broad a view will you then have when seeing it throughout the entire Universe.

Chapter 20

Practical Application of Reason

Practical application is something that is not well understood by most people. Most people go to school or get a job with a specific task or occupation and that is all that they ever learn throughout their entire lives, and some proceed to argue for or against the Bible and science. The quantity of holes in their theory explanations is often quite substantial. Of course, there is nothing wrong with only doing one occupation, but you must understand that when you only know how to do complex mathematics, then you only know how to do complex mathematics. If you lack the ability to apply that math to substance and reality, then you really do not know mathematics as you believe you do. This goes back to the methodic learning section discussing learning via *function*.

Most of us believe that we understand our belief well, until holes begin to appear in our understanding of our belief. Then to fill in those holes we begin to accept nonsensical theories. Often these theories are far-reaching and stand outside of reasonable explanation.

We can reason through just about anything with irrational unsubstantiated thoughts, but this won't make those irrational unsubstantiated thoughts a reality. The same practical application method that is used by those who utilize *functional* thinking must be used if we want to avoid irrational thinking.

Your Freethinking and Reason

We live in an era where free-thinking is interpreted as a sort of rebellion against the status quo. This is an age-old problem that we cannot quite seem to pull humanity free from. It might seem to be a good thing that people would rebel against the status quo and think freely, because the status quo is often wrong. But often wrong is somewhat of a relative term. Yes, the status quo is wrong now and then, and this is true all too often. But usually, we have things right which is where "free-thinking" fails in that case.

I deeply believe that people should think freely and not be tied down with incorrect beliefs, but if you do this without utilizing the seeing-through-function method of thinking and reasoning, then you really are likely not thinking freely. If you have been told by someone who appears "smart" something that seems logical, and you decide that you want to break away from "that old-time religion" and believe this smart person's ideas and then proceed to internalize those as a "free-thinking" deeper belief, then you may have been duped. True free-thinking will never blindly believe what someone else says without first testing those thoughts through the methods of function. True free-thinking is about abandoning lies and then seeking only Truth.

You might be able to dream up your own "free-thinking" ideas while gazing at the heavens, but again if you fail to use the means of *function* when considering the ideas that you have dreamt up, and if you are afraid to have those ideas challenged, then you are not really thinking freely. Instead, you are only imaging things *that you want to believe* because you like your own ideas.

The "reason" aspect of the free-thinking ideology is also quite misunderstood. What really is "reason"? "Reason" is when you take a thought that was developed by someone who was thinking freely and you then look at all of the aspects of the thought to see if those aspects could produce the result implied by the thought. The method of "reason" is desirous of using function, but if it fails to do so, then it is not really *reason* or *free thinking.*

Knowledge is Power... But...

There is this phrase that says "knowledge is power", but this belief is dangerous because what we *believe* to be knowledge may in fact be a lie. What is *knowledge*? As is implied by the fact that the word knowledge contains the word "know", knowledge is the culmination of what you know when you have "knowledge". But that "knowledge" is not really always true, and if it is not true then it is not truly **know**-ledge.

It is not possible to "know" something that is not true. You can know that the thing is not true, but you cannot know the untrue thing because it does not exist. So to be "knowledgeable", you must be accurate and true. Yet too often, we believe questionable speculations as if they are fact. This causes us to have the erred thought that a person with incorrect "knowledge" is "knowledgeable", when in truth they are not.

Many people "know" incorrect things and build their entire lives around the incorrect things or thoughts, but if those thoughts are incorrect then you are only a believer and you really have no knowledge. But don't lose heart if you currently believe wrong information, because that can be turned around quickly when you accept the errors in the thoughts as the errors that they are, because now you *know* they are errors. That minor distinction is really quite important.

You can either believe you know something and be wrong, or you can test the thought through the utilization of thinking through function and realize the errors in your thoughts. Then your pseudo-knowledge about something incorrect can become actual knowledge that the "something" *is* incorrect. By nothing more than a simple decision and a small amount of functional thinking, you can go from being very wrong to being very correct.

The Scientific Method

Let's examine this ambiguous term "The Scientific Method" and see if we can come up with a consensus as to what "The Scientific Method" means. The scientific method is supposed to be an approach to science where through observation of a function we conceive a hypothesis and set out to create experiments to verify if our hypothesis is plausible. However, there is often more than one possibility of what could have occurred.

Consider the freezing point of water for instance: Because salt water freezes at a lower temperature, we could hypothesize that inland creeks carry saltwater and that is why creeks are not frozen when the lake is froze over. But while salt content is one reason water freezes at a lower temperature, water also will not freeze in a flowing creek even though it is below 32 degrees F, or 0 degrees C.

We must take great care in not allowing ourselves to get caught up in false or badly flawed hypotheses if we ever hope to advance true scientific knowledge. With regard to Creation topics, perhaps the more heinous is the foolish attempt to combine six-twenty-four-hour-day creation with big bang. Adding two very incorrect hypotheses together does not make it a correct one. Rather, it increases the likelihood of not being able to escape for the believer of such foolhardy theories.

Believing In versus Believing That

We can believe that the Creator exists, but we can also believe *in* the Creator, and while these two ways of believing are very similar, they are also different. We can look at this in terms of people for instance. Consider a capable person who you know can do some task because they have the strength and skill to do the task, but you happen to know that the person is not very reliable and you know they would fail at carrying out the task. You believe that they are capable, but you do not believe *in* them personally to a point where you know they will do it, and you know this because they lack a certain kind of integrity and have proven unreliable.

Believing *in* versus believing *that* is mostly the difference between tangible and intangible. When we believe *in* something then in our minds we give it special qualities of integrity—it is a personal sort of connection. However, when we believe *that*, then it is more viewed as a functional action or is tangible. It is a matter of trust: *believing in* is to trust the person; *believing that* is to trust that the information is true

But to *believe in* and *believe that* are both a matter of whatever you have accepted as being true regardless of why you accepted it. Even if you are willing to change your mind, it is still only a belief if it is not **know**ledge.

We can know something and *not* believe it. Or we can know something and believe it too. *Believing* is not *knowing* and *knowing* is not *believing*. We make our convictions based upon our beliefs, not based upon our knowledge. Many people refuse to admit to this, but it is true nonetheless. It is our belief that drives us to test our hypotheses.

We can test to disprove something and claim that we are not believing that which we are attempting to disprove, and it is true that we do not believe that which we are working to disprove, but then what we believe is that it is not true. All science is based upon beliefs and trying to prove whether those beliefs are true or not.

Regarding believing in a Creator it is different because we are going to choose to believe without specific evidence and it is more a matter of personal experience, connection, and issues regarding trust in that Creator. But the logic we use to reason through that belief is often lacking and follows falsehoods, when it really does not need to be that way.

Scientific Creation Predictions

One area of science that is often used in somewhat of an overreach is that of "prediction". Scientific predictions are very helpful when attempting to assess a theory. In basic physics this is easy to do because we can do it by making predictions and then follow those predictions doing simple experiments to test the predictions in a lab. But with astrophysics it's different because we can't create stars or planets to test things, and we have a difficult time with fusion due to the dangers involved, leaving us to create theories to test our theories.

Take the WMAP space temperature probe as an example. The WMAP is more than just a theory because it is real piece of equipment that was put into orbit that sends back real scientific data. However, while the equipment is real, and it appears that the temperature range seems to nominally differentiate in the various regions and therefore exists, it is still only theory or simple speculation regarding **how** those temperature differences exist.

Any atomic energy occurring in space at the scale that space is, and with the quantity of stars and galaxies in that space, is likely going to have some residual energy. Big bang theory claims that if the bang stretched out via expansion as expressed in the theory, then nominal amounts of residual energy would still exist. The WMAP probe was sent into space to test this theory. However, there are several problems with the premise of this aspect of the big bang theory, which is that almost all creation event theories would have to make the same claim if they are being scientific about the physics of actual Creation.

The thermodynamics and the laws of physics require that some residual energy would exist with activity on a large scale that is the size of the Universe. But additionally, we also must realize that there is little information available about the WMAPs repeatability, and more importantly, what it is actually reading. At this point we are not questioning if the WMAP was reading residual energy, but rather what the *cause* of the energy was or is.

So then, we have to go onto question what energy specifically is WMAP actually sensing? And, are the readings even *repeatably* accurate to begin with? As mentioned in earlier parts of this book, the temperature differences sensed by WMAP are very minute. Without a high level of accuracy in repeatability, that data simply cannot be trusted. We cannot be certain if the energy being sensed was residual or if it is newly generated energy such as the energy from distant stars and galaxies reaching the WMAP probe.

As you can see, scientific predictions can and should be made, but when the predicted circumstance is found to exist then we have to ask further—what else could have caused that predicted circumstance?

Predictions are only as good as the conclusion. For instance, you could wake up from a dream that you are a trillionaire and that various banks were holding your fortune. So you theorize that if you check the holdings of the banks it will add up to the amount of your dreamed liquid-financial-worth. You then set forth to prove your theory by questioning the banks involved and you find that the trillions do in fact exist. At this point you now believe that you have proved your theory that you are a trillionaire, but in truth you have only proved that the holdings of the banks exist. On the other hand, the cause of the banks' holdings is an entirely different story. Since it is unlikely that you have made any more than maybe a few hundred thousand dollars per year for your entire lifetime, the aspect of your theory claiming the trillions to be *yours* is a bit of an unrealistic reach even though the trillions do in fact exist.

Even if you earned a billion dollars per year it would take you one-thousand years to save a single trillion if you spent none during the one-thousand years. So your theory, while it is accurate in assessment that the banks have large holdings, it is not accurate or realistic about the trillions being owned by you. We have to guard ourselves in this respect, and make sure to not overreach on what a scientific prediction actually means.

If a Creator Created, then we should be able to make some predictions, but if we want credibility in doing so, then those predictions should have a more sound foundation than do conclusions made regarding the WMAP results. If a Creator Created, then our first point of interest is the "heaven" and "earth" spoken of in the first sentence of Genesis One. What are "heaven" and "earth"? We have already established that they were likely not the "Heaven" with the stars and other celestial matter, and not the "Earth" that we live on at that early point in the Creation account, that is to say the first sentence of Genesis One. So, what are they?

The second point of interest is: when light was made to be, then what action took place regarding the pre-matter-earth? The third point is what was/is "firmament"? The fourth point is what were the waters that were "divided"? And fifth, what manner of division occurred?

Once we establish what we theorize was involved in each of these points of interest and the specific order in which they took place in Creation, only then can we even begin to make any sort of intelligible predictions regarding what might have occurred and what we might now find as evidence when trying to understand Genesis One.

The problem we have regarding making such predictions is that it is far too easy to take what we know today, which is both very much and very little, and then we try to guide our theories towards that end. Where if we did not know what we believe we "know" today, then we would possibly conclude or theorize to a completely different end. So, our first human hurdle to overcome is that of our biases and obsession in directing our theories to our desired-predetermined outcome.

The way to approach the problem is to first ask questions about each step and see if each of those questions can be answered with the Genesis One text without overreaching or making things up that don't really answer any questions and are not actually stated in the text. Adding things to the text that it does not indicate and that could not occur based upon the stated order of events is why Christians always lose the debate when discussing Creation versus big bang. It does not need to be this way.

Chapter 21

It's Our Own Responsibility

Whether you are an atheist big bang believer or a Christian Creation believer, you have a responsibility to yourself to acquire *accurate* information. We humans have a tendency to point to others as the responsible party. In other words, "it's not my fault, that's what I was told", or "I can't help it because of my DNA" etc. We try to place the problem and/or blame elsewhere so that we don't have to answer the tough questions for which we are not prepared to answer.

This shifting of blame is a much larger subject than just our own personal shortcomings. We do this in science on a regular basis, and I am not referring to blaming others for our own lack of understanding.

Far Away Things

The Universe is quite obviously a vast topic covering a vast area, and as mentioned in a previous chapter, it would take very long using our current rockets to even get to the nearest star. But we will likely not get any sort of ship or probe to another star in our lifetime because we always assume that things are other people's responsibility. If your quest is to find a way to travel at tens of thousands of miles per second, but you believe that someone else is already trying to do that, so you completely abandon the idea, then you have decided that it is someone else's responsibility.

Distant Remote Control

If someone did not believe or theorize that we could possibly explore Mars remotely from here on Earth, then we would not have sent a mission to Mars for exploration, or even tried to do so for that matter. This also was the case when people first set foot on our Moon.

This holds true for interstellar travel and communication as well. If we are only ever able to achieve the speed of light then it is highly unlikely that we will ever get anywhere near another star. And unless we can find a way to communicate at a speed far greater than the speed of light, it would take about eight years to receive an answer back from a space probe from the time we sent the request, when the probe reaches the star nearest to our solar system. This basically means that it would be almost impossible to control the interstellar probe because time between each command would need to be a total of about eight years as we await the response of each previous command sent. That is four years for the probe to receive our command, and then another four years for us to receive the probe's return response at the speed of light.

Here or There

Whether here or there, the actions we take are going to be similar, but the time involved changes dramatically with distance. Because we have believed incorrect information about mass, velocity, and the constancy of light's speed, we have chosen to ignore our inquisitive nature regarding interstellar travel. And the imagining that we actually do decide to do regarding interstellar travel is always based on nonsensical fantasies such as wormholes and warping space. We base our time altering *fantasies* on the errors about supposedly not being able to exceed the speed of light.

No amount of irrational thinking will get us beyond the far reaches of our solar system. Science is Science and Physics is Physics, and everything that any of us has ever experienced of a tangible nature abides by the physical realm and is controlled by the forces that we describe with our "laws of physics", which have no regard for Einstein's speed limit on light.

To unlock our minds to be able to further study God's science, we must first free ourselves from the erred thinking that all of us at times take part in.

If this Creator God does in fact exist, it seems certain that our ability to explore the Created is at the top of the list of important things to do to get to better know the same said Creator. And that is the point of our existence—to know the Creator.

Chapter 22

Questions

Questions are a peculiar and actually very interesting topic. What are these words and thoughts that we assemble that cause us to gain information?

Our questions have a few steps or parts that we need to do to arrive at a fully complete question, and without all of those parts being done, we will most likely arrive at incorrect conclusions if we arrive at anything at all.

Formation

The very first step is the initial formation of the thought that provokes the steps that will follow. What causes us to arrive at the first step?

When we experience a gap in our understanding we then begin what is usually the unintended process of wondering about that gap. We humans are conceived and then we are born into a world that is ready to go. As newborn infants, we do not need to seek our food because it is typically brought to us. Yet we do seek. If you have ever touched the cheek of a new-born child you likely noticed that they will typically turn their head towards your finger in attempt to find their food in the same way they seek when they are being held by their mother as she positions herself to feed the infant. This eating response is not as strong and will sometimes not work once that child has been fully fed. This wonderful instinctive nature to seek is what allows us to live. Without our ability to seek we would quickly die. I feel quite confident that most people do not realize the amount of questions they typically ask themselves in a single day. Those questions begin the moment you open your eyes in the morning as you wake. But in truth, they are also being processed all night long as you sleep.

As we navigate our way through each day, we encounter more input than we care to conceive. Every single color, shape, texture, smell, taste, and sound, etc. invokes in us a tiny point of information gap. Once we have become familiar with those information gaps, we quickly find the answers and we then become proficient at filling the gaps at a pace that is really quite astounding.

You will notice that children are very inquisitive as they wander around the house and explore. They do this to fill in the gaps. Regarding children and newborns, the gaps I am referring to are seemingly insignificant to us adults. As adults, we question and then answer so quickly and efficiently that we generally don't realize that this is occurring.

Children wander the house in a seemingly aimless manner, but it is generally not as aimless as it appears to us adults. As adults, we typically have a specific purpose when we do things. But since children do not have any specific tasks that they are required to do, they have the luxury of endlessly exploring. Anything that catches their attention will be investigated by them.

Your Attention Please

What causes something to grab our attention? The stimuli we receive allow us to notice something. We notice things typically because they are out of the ordinary and thus to us they stand out. "Out of the ordinary" can be anything that arouses our senses and it is somewhat dependent upon each individual person. We see, hear, taste, touch, and smell things as a continuous occurrence, and it is when something stands out to us as unique that we will then explore; much like when a young man is looking to find a girlfriend, he will see many girls, but only some girls will catch his attention and vice versa.

When an item grabs our attention, at the most base level we immediately question simply "What?". It is a sort of alerting call our mind does in preparation to explore. When the "What?" spark is ignited in us, we begin the pursuit to satisfy that initial "What?" This **attention please** part of our ability to question gets us in much trouble when we are not alert. In our pop-science culture people are constantly screaming for our attention and for us to believe their own theories.

The alerting of our senses causes us to first become aware of something causing us to ask "What?"

Our Wonder

The next step of a question is to wonder. The "*what?*" part gets our attention, but it is our *wonder* that is the beginning of our pursuit. After we have been alerted, our mind has an inherent desire to investigate, but this will depend upon our moment-by-moment analysis of the **what** that ignited our quest. We will seek answers regarding the subject of our "What?" and when those answer gaps are filled, we then move on to something else.

Our *what* generally turns to *wonder* nearly immediately and could be thought to be almost the same thing, but it really is not.

Wonder is our ability to give thought to something and to process our **what**.

Seek the Quest

As we wonder, we set out to seek to fill those gaps about which we are wondering. Our quest is the seeking part that we generally see as the *asking* or the *questioning*, and in a way, it is, but the seeking I am referring to is more internal within us. Seeking is more about us working to form the question in our mind.

You might think that forming a question is simple, you just look at something and ask about it. Right? Not exactly. Forming a question is a skill that we all have and we developed at a very young age. Then as we age, what we seek becomes more complex, but sadly this can cause us to lose our desire to know more details about the source of our *what*.

As we age, we experience a great deal of discouragement when seeking or asking about things. This begins when we are very young and mostly comes from parents and teachers, but also from siblings and friends who might be a little older than us and also from society and even our peers.

If you ask a question that doesn't make sense to others, then a typical result is that they will ignore, mock, or become frustrated with your questions. This usually results in us withdrawing, and all too often, we then abandon our quest altogether. After enough of this we begin to form a bad habit of giving up before we even get started.

You can observe this regarding the topic of how things came to be. There are many books and video programs that we can use that appear to have it all figured out. And depending upon our level of interest, which is usually determined by everything previously mentioned here, we might decide that our information gap has been filled by some of those programs. Or maybe, since the content of the program is hard to understand, we internally wrongly think that somehow *we* are not intelligent enough to understand it all.

The Question Why?

If we manage to overcome our discouragement, then we begin to form the questions for which we have been seeking answers to fill the information gaps of our initial **what**, and those questions will be in some form of *what, where, when, why, who,* and *how.*

Our ability to allow our gods-of-the-gaps in both religion and big bang science fails our ability to question, and that is our cop-out used to avoid answering the tough questions for things we do not understand. We do this because we want to hold on too tightly to our pre-concepts of what we already believe are our answers.

We take Einstein's theories as if they are fact, and in doing so we make erred assumptions that stop us from further questioning and researching areas his theories address.

If we read Genesis One in a scientific manner we should be able to make some predictions and then we can seek out those predictions to see if it is possible that they might hold true.

Consider the quest for "exoplanets", that are satellites of other stars. In science, we have acted as if this is something amazing and that the chances of it occurring has some sort of slim odds. But if you read Genesis One in an authoritative Bible version and do so from a scientific perspective, then you can come to no other conclusion than that all stars likely have planets orbiting them, and further that it would be unusual to not find them. It is more likely a very rare event that a star is void of planets. Thus, we can predict that given the proper equipment and enough time we will find that half, and likely far more, of the observable stars have planets. And I use only the term half because if the exoplanet orbit from our view does not cross in front of the star then detection is far less likely.

Based upon a scientific analysis of Genesis One, we can also predict that light's speed is not constant and that light is bent or warped rather than space being warped. Light may well be constant the moment light is emitted, but it is likely not constant from one emission to the next emission whenever the speed of the emitting body has changed relative to its first emission of light.

It would be a needle-in-the-haystack sort of quest and generally not possible, but if we could find a beam of light being emitted on a spinning body out in space, and if we were able to see the emitted radiation, we would likely see beams of light that were curved or arched as the spinning body rotates. The curve would not be due to anything other than the light leaving a trail as it is being emitted. Once emitted it will also likely continue move in the direction of rotation from which it was emitted as well as the direction at which the emission was directed much like a spinning firework.

And from a practical application approach, since light's moment of inertia is so low and the "moment" is very small, light would have a result of travel at a slower sideways velocity than the star is rotating at. This would be different than if you were to throw a rock as if it was a light beam because the moment of inertia of light is much different than the moment of inertia of you throwing a rock in the same direction as the light beam is pointing.

We can make predictions on any data, and often those predictions are wrong, but we can also make predictions that are very accurate. When we fail to question, we will generally make no predictions at all. Our predictions are our first attempt to fill in our information gaps, and our testing is intended to prove those predictions or theories, thus completing our job.

Our information gaps are our gold, and our ability to fill those gaps can end up being a crown of glory, or the hat of a court jester. At some point we must decide which by making a choice as to whether or not we will *properly* fill our gaps.

Chapter 23

It's a Matter of Choice

When we set out to make our predictions it should be done in an honest manner doing our best to banish our human prejudices regarding what might have occurred, the sequence in which it occurred, and what could have initiated it.

This means that we have to choose to be honest in *all* of our assessments. In the New and Old Testaments this sort of honesty was referenced indirectly whenever the point was being made that we should not judge people because of their financial or social status, but instead, and punishment aside, we are to judge what is in their heart. The same is true here; we each must examine our own heart and mind before theorizing, in order to make sure that we don't force our desired end into an otherwise plausible theory.

Relying on Someone Else

Many people who are involved in the creation/big bang subject are schooled in a manner that is utterly opposed to any sort of discerning Creator and will vehemently promote that viewpoint, thus following suit with their education. You cannot be considered honest in any assessment when you disregard any viable possibility.

The same is true for six-twenty-four-hour-day creationists who are schooled by preachers who do not understand physics. Creation is only a viable possibility when the text of Genesis One is not perverted by translators who have grossly misinterpreted the text and rewritten it for all to see. What we all need to understand is that if this Creator exists and Created Creation, the laws of physics that we humans invented have only been invented because they can reasonably accurately describe various aspects of Creation. And since those laws work so well for us in space exploration, we can make the relatively safe assumption that those laws apply across the entire Universe.

But relying on someone else's mathematical calculations means that you are relying on their expertise, or lack of thereof, for your analysis of the situation or the question at hand. In other words, when you believe in other people and their math or assumptions without you verifying it all, then you have made them your god and you serve them because you are going to make all your decisions based upon their assessments of things—they have become your religion and your god. This is not speaking of the laws of physics, but rather the improper application thereof.

The laws of physics that we have created may have some slight inaccuracies in the calculations requiring slight adjustments to be made. However, those laws are accurate enough to enable us to send probes to Mars and other planets and calculate reasonably accurate arrival and landing information. These laws are sufficient for landing and operating a remote vehicle to enable us to do some exploring and report the findings back to Earth. But that ability is built on very consistent laws which are built on very consistent theories and vice versa.

Inconsistent Imaginings

Many atheists imagine that the Bible is inconsistent with everything that we see and witness, but let us put to the same test the theories of evolution and big bang. Do not imagine that there is no evidence supporting the Bible's view and that there is only evidence supporting the big bang view.

With Creation, the view is that the world was made in six-twenty-hour-days, yet if you actually take the time to read and then analyze the Genesis One text *and the data* without prejudicial biases, you will then see that "heaven" and "earth" were made, and later after that, then the first day occurred. Do I have my thoughts on this? I do and with regard to this book you should have a fair idea of what they are, yet it is not for me to dictate to you what to believe or deduce whether it agrees with my current view or not. The purpose of this book is to reduce the blind faith that too many people have with regard to science and also with regard to an erred scientific view of the Biblical side of science.

Many people are tired of listening to all of the scientific pontification of inaccurate theories and equally as tired of similar misinterpretation of the Bible. It is odd that someone can actually read the text of the book we call Genesis and then insist that our Earth was made within six twenty-four-hour days and that light was created, and then go on to tell us that Genesis does not make sense because the Sun was not yet made when the light was there. To which I ask, what has the Sun to do with light?

Reading this book up to this point should have offered a clear view that it is more leaning towards pointing out what did not occur rather than specifically what did occur.

A foundational aspect of Creation is the Creator, and when using scientific big bang logic, the Creator did not exist and therefore the Earth came from the big bang and thus there is no proof of a Creator. Is it a wrong approach to say that there is Creation therefore there is a Creator? From a Creation perspective, no, it is completely logical. However, if you do not believe in a Creator then you obviously cannot use that approach and you cannot use Genesis One, leaving us in a position of not being able to explain a lot.

Our human logic seeks a *causal* result. In other words, we are here, therefore we were somehow made. In science it is through "natural causes", meaning that something or some cause without conscious caused us to be here. This is no different than the "there is creation therefore there is a creator" approach. In both cases you are making some cause to be the method by which we got here.

In the erred scientific approach of big bang it is done through incrementalization over long periods of time, or it is done so instantaneously that it happened before the laws of physics had a chance to get going. Where with the standard twentieth century six-twenty-four-hour-day view of creation, it is excused away with—God did it all with a hocus-pocus-like POOF! Both of these erred viewpoints must be abandoned and the entire subject should be reinvestigated from scratch utilizing our scientific equipment and a fresh *unbiased* view of the Bible using older authoritative versions.

Anecdotal Evidence

When someone dies on an operating table and sees themselves from the outside of their body and then later is revived and reports on this occurrence, then it is called "anecdotal" even though this has been experienced many times and those who experienced this can describe things that they should not be able to describe that occurred during their "death" period. Such a phenomenon cannot easily be scientifically tested because this sort of event typically occurs at random on the operating table and the medical staff is busy doing everything they can to bring the person back to life. So, what really is enough evidence for "science" to make a conclusion? When is it not "anecdotal"?

What happens if we apply the same standards to big bang as we do the death on the operating table situations for instance? By the same standards, is not big bang anecdotal or even worse speculatively-anecdotal?

When a person who is highly regarded as a "credible citizen" relays their account of an "out of body experience" on the operating table, their *firsthand* account is apparently not credible. But when some big bang zealot pontificates on their theories of big bang and the science behind it all, it is somehow regarded as "scientific fact".

"Anecdotal" means that it is not necessarily reliable or true because it is based on a personal account rather than being based on study and facts.

So, if the doctors state a fact that someone died on the operating table, that point would generally not be disputed, but the person's account of what they believe they experienced during that period typically is disputed. Regarding big bang, if you disagree then you are cast as some sort of loony creationist, yet big bang is not based on accurate research at all, and nearly all "evidence" for big bang is greatly perverted or manipulated and extrapolated to a point that it can no longer be considered "scientific", and as such becomes beyond "anecdotal" to a point of utter dishonestly.

If something such as the WMAP mentioned earlier is "evidence" of a big bang, then it can only be anecdotal because it is only evidence of such based upon an interpretation and belief by each person, and that person relays their far-reaching conclusion to others anecdotally. Their personal interpretation or experience regarding the WMAP is "anecdotal". This is true of nearly everything regarding the religion of big bang.

Your Fears versus Your Rejecting Information

When discussing God or a Creator with many self-proclaimed atheists, it is clear that there is a great fear of the afterlife. In other words, many people fear hellfire and so they choose to believe that God does not exist. But also, we have some creationists who have not learned to think through the science of things and will foolishly reject science because in their denial they foolishly believe that the only viable scientific option is the big bang explanation of creation. They reject that because it does not fit with their desired six-twenty-four-hour-day creation beliefs. Yet there are Christians who want to understand science, but will blend the two or do whatever they can so as to not outright reject God for fear of hellfire.

There's a difference between fearing something and rejecting something. When we reject out of fear, then we are not really rejecting the information, but rather are choosing a path that is emotionally less painful for us to bear. When we reject out of pure logic then we have a real basis for our rejection. For instance, big bang can be rejected solely on the basis that the laws of physics that make the math of big bang work must be nullified momentarily in order for big bang to be able to occur.

Be Careful what You Choose

Be careful what you choose to believe because you have to live with the actual consequences. If we believe in God and we are wrong then we are only fools. And if we believe in big bang instead of God then we might end in eternal hellfire. If we believe in a Creator and we are right then it will likely be credited to us.

What we have to understand about a Creator, Creation and an afterlife, or about big bang is that it doesn't matter what we choose to believe because it will not and it cannot change what is real and true. Regardless of how afraid we are of an afterlife or how afraid we are to feel humiliation for changing our belief path, we will still always have to deal with the consequences regardless of what those consequences are.

If there really is a God and we deny that Truth, then we will have cheated ourselves out of what sounds to be a very good deal. But someone who is a six-twenty-four-hour-day creationist and has a deep faith in God may also be in a similar danger as those who deny God. How could this be?

If you believe in six-twenty-four-hour-day creation because your "Bible says so", but your favored Bible version and you happen to both be wrong, and then because of your gross misinterpretation of Genesis One you then go on to pontificate your error-filled interpretation as "factual" thus causing other people turn away from the Bible and God because what you are telling them about the Bible is simply not logical or scientifically viable, then are you not somewhat responsible for those souls who have followed your instruction? Will you not have some sort of price to pay for being a "false teacher"?

Whether it be "Science" or "Religion", beware of the false preachers, for there are many who will deceive. This does not mean that they mean any malice or harm, but rather only that they are wrong. It is generally easy to spot those who are malicious and want to lead you astray from what is Truth, and while those few are vile disgusting human beings, it is the people who are true believers in their religion or their science-religion that are the most influential to you and therefore truly the most dangerous to you when you believe them and they are *incorrect*.

When someone is deliberately leading us wrongly we can usually detect it and we will reject them. But when someone is passionate about their belief, it is more compelling to us thus causing us to more readily be convinced about the chosen religion or science-religion. When we find a credible person who says what our itching ears want to hear, we have a strong tendency to believe and follow them, rather than using basic logic and reason.

This is only bad when they are wrong, but it is always dangerous to us because if they take a wrong turn then we go down the wrong path with them because of our blind faith in them and their erred theories.

Be careful what you choose to believe, because it will affect your subsequent logic and thoughts and can lead you on a very inaccurate path of folly.

Chapter 24

Infinite and Eternal

It is interesting to note that in a later part of the Bible it talks about a new heaven and new earth. We could couple this with what we believe about science and thermodynamics and how long the Universe might last. I am not indicating that the reference to new heaven and new earth must be used scientifically, but it is an interesting point to consider.

For our human purposes, *infinite* and *eternal* are good descriptions of "God" or, better stated, the Creator. But do we really know if the Creator is infinite or eternal? No, not really other than what it says about that in the Bible. Does it matter? No, it doesn't, not for the scientific aspects of Creation anyway.

Infinite and *eternal* are different but similar, though often interchanged. Infinite tends to be more indicative of size with regard to Creation, and eternal tends to be more associated with the idea of duration or length of time.

Scientifically, we believe we know that the Universe may be limited in life, and therefore it is not eternal and people on all sides of the big bang versus Creation debate agree that there was actually a beginning. Though there are those who will invoke the cyclical big bang theory that everything has always been contracting and then once again big banging and then contracting again in an endless cycle every handful of billions of years, but, that belief is held by very few rational minds.

For Creationists there is only one distinct beginning and it is the Bible's Creation account. This is true regardless of what the chosen time-period or order of events was. The Bible doesn't discuss any sort of multiple beginnings of the past, so "cyclical big bang" theory is not a part of the Creation side of the debate.

Our human logic tells us that there was a beginning and that it was somehow initiated, and most people on all sides of the debate tend to feel that the Universe is not eternal.

An infinite Universe is another story though. Big bang theory indicates that the Universe started at a single point in all of space and then exploded and, depending upon the theory, is still expanding and is therefore finite in size. But then to fill in for the mathematical anomaly that particular point of view creates, they will pose multiple different locations big bangs occurring throughout the entire area of space.

Creationists don't really get much into the infinite issue with regard to space, which is probably best because there really is no way of knowing that, and to speculate on it is always going to be somewhat of an overreach. However, Creation by a Creator does allow for the possibility of an infinite space and a Creation with no physical end.

Again, it doesn't really matter because we can only detect what we can see with our telescopes and we cannot prove or disprove if gravity has its limits regarding distance that might inhibit any of our theories.

Do Unto Others

As mentioned in the last chapter, if we lead people astray we eventually will either face embarrassment or hellfire or both. Jesus said "do unto others as you would have them do unto you", meaning that you should treat others as you would like them to treat you. If you do not want to be led astray in your beliefs, then you should not lead others astray in theirs. The hidden problem here is in our humanity and the way we use our logic. If you do not use sound reasoning and logic, then you will most certainly be led astray at some point in time.

What all people, especially six-twenty-four-hour-day creationists must bear in mind, is that the Creator in no way would want to lead us astray, and the Bible indicates such, as it is very informative about the result of the people following bad influences. The Creator, if real, is going to want us to *accurately* know and understand Creation through using the sound logic and reasoning of our god-like nature, as in "created in the image of". Anyone who believes that we must believe the Bible based on faith alone and dispose of all logic, scientific or otherwise, is a fool–It is quite the opposite, we are to use our God-given logic to discern these things.

In the Bible's Matthew 7:1 it says something to the order of "judge not because with the same measure you deal out to others it will be dealt out to you." This is where the fear of hellfire can come in for many people. When we do wrong, we have this inherent understanding that we will pay a price. What we fail to understand is that when we harshly judge someone we will be judged harshly also. So, if you judge someone harshly and unfairly for being wrong, then beware when you are wrong. And the more wrong you are then the more harshly you could be judged–Wrong is wrong, but your wrongness is increased in this case by how many people you lead astray.

This is consistent with an "eye for an eye" which is a warning to us that what we do will come back on us. We need to guard ourselves against our own prejudices so as to not lead others astray, because we will inevitably be led astray by others when we use prejudicial reasoning.

The Universe Does Not Abide by Our Rules

If we want to understand Creation done by a Creator, then it is imperative that we use scientific logic—it is ***not*** an option. Someone serious about the Bible's Creation account has a much easier task than does someone who rejects the Creation account entirely; this is because if you reject the Bible's Creation account then you are rejecting good and useful scientific sequencing and are attempting to force science and thus ignoring facts and data in order to avoid anything that might agree with the Bible's account of events, thus leaving you in a compromised position. On the other hand, if you are utilizing the Bible's Creation account *along with* real science, then you are allowed to use that information in harmony with science.

But do not misunderstand this, I do not mean that someone should believe any scientific pontification that comes along. What I am getting at is that we must use logic whenever discussing Creation. Because if there is a Creator, then basic human logic should scientifically confirm the Genesis One account of Creation since the laws of physics are a result of the Created matter. Any other path of logic is foolish when you are trying to scientifically confirm the Genesis One account of Creation.

If Creation was done by a deliberate Creator then everything that we experience here on our Earth or in the Heavens will abide by the forces that dictate our human-written "laws of physics". Choosing to disregard any of these laws that describe so much of what we are beginning to understand is foolish. This does not contradict what I spoke of regarding the laws of physics in *Bending the Ruler – Time Travel, The Speed of Light, Gravity, and The Big Bang*. In *Bending the Ruler* where the point was made that the laws of physics do not dictate the forces, but rather those laws *describe* the forces–Nature is not subject to us, we are subject to nature.

The Universe does not abide by our rules or laws, we abide by the Truth and the Order of the Universe and its forces and there appears to be no avoiding this simple truth. If there is a Creator that made all things, then that Creator is the most logical scientist that ever existed, and everything that was made would likely have occurred through that logic. Logic and reasoning are not some random aspect of humanity owned solely by "science". Logic and reasoning are the sole means by which we can best know this Creator. Some would say that love is the best way, and while this may be true, logic and reasoning are very important aspects of the love we all understand.

For well over a century, since the abundant ability to inexpensively reproduce modified versions of the Bible, our wonderful human logic has been eroded, thus allowing vast misinterpretation of the account of Creation detailed in Genesis One. I am not sure who is more to blame; is it the "scientists" for not noticing and pointing out the gross errors written into many altered Bible versions of recent centuries? Or is it the people who translated those altered versions? Or is it anyone who believes those versions to be accurate representations of what was actually originally written.

Any self-respecting scientist who has in any way ever studied the Creation aspects of Genesis should have picked up on these translation errors in Genesis One and pointed them out to the world. But as our human nature would have it, we instead get confused as to which version should be looked at, and then we become frustrated and altogether abandon the Bible and make a decision that the Creation account is just a "fabricated story" inscribed and altered to suit our beliefs. This is both a horrible lie and a terrible truth.

As explained in *Understanding The Bible - The Bible How-To Manual* AND *The Things We Don't See*, all Bibles are not created equal, and there are versions that are authoritative and there are versions that are *not* authoritative. The older authoritative unadulterated Genesis versions describe a very compelling and basic scientific account of Creation, thus making any thought of the Genesis account having been a fabricated story an outright lie. But on the other hand, the Bible has been altered to suit our beliefs with those newer adulterated versions, making that point a terrible truth.

This is not intended as some sort of deliberate condemnation of anyone's chosen religion or beliefs whether it is *religion* in general, or the *religion of big bang*. The intention is for all of us to reexamine our thinking and reevaluate our scientific findings, and especially reevaluate what we *think* the Bible says.

Infinity

Is there proof that the Creator exists? That is all going to depend upon who you ask. If you ask a staunch atheist then there is likely little or nothing that will change their mind because it is already made up and the *belief* that there is no God is *their religion*. They only have the flawed explanations of science to explain how everything got here.

For those who believe in any form of Creation by a Creator the evidence is everywhere, including yourself and the ground that you walk on.

But when trying to logic through the question of big bang versus six-twenty-four-hour-day creation with our human reasoning, it really cannot be done because the scientific flaws in both ideas are vast and do not match with what we know to be true about physics or reality. Both are illogical and dishonest.

When we use our human logic, this whole Creation discussion will always come down to the very initiation point of Creation.

Big bang fails because our human logic will always look for a beginning. So, to compensate we cycle big bang from singularity to fully expanded and then back to singularity again, over and over in an endless cycle. While big bang tries to posit a beginning, we must oscillate our mind in opposition to singularity so as to avoid the underlying truth. Big bang fails because even a single point of theoretical non-existence is *something*, and the question of how it got there is avoided entirely. Even if something was in existence infinitely back, we still have the question of how did it get there? This logic is somehow born into us and, though we try, we cannot avoid that logic of the initial beginning as is eventually proven in *any* scientific rationale we ever see.

For those who believe that everything was Created by a Creator, the beginning is explained–it was done by the Creator. Leaving only the question of how the Creator came to be. But at some point, something had to come from nothing, and scientifically speaking, since as far as we can scientifically prove, our souls/spirits are not tangible and therefore are "nothing". So, the origin of the conscious discerning Creator falls outside of the discussion of this book as this book is about the scientific order of Creation as stated in authoritative versions of the Bible.

When accepting the concept of a *Creator* that is all spirit and can't be directly scientifically detected with any instrumentation, the question still remains: How did the first of parts of the Created come to be? By what method did this Creator Create the pre-matter "heaven" and "earth"?

If you imagine the darkness of space without any light or anything in it, but rather only a vast nothingness and then just think about it and how everything came to be for a while, you will be forced to grapple with something just spontaneously appearing without some sort of cause. Big bang does not address that point and our logic drives us to demand an explanation. Creation offers that logic and explanation, but it is often contaminated by those poorly translated versions of Genesis.

Keep these various issues separate when pondering this topic. The astrophysics aspects of Creation have nothing to do with evolution, so evolution is not part of this discussion. Also, big bang is not science and science is not big bang. The pop-culture scientists do not own science, and most of them are wrong in much of what they say and believe. Most Creationists do not believe in six-twenty-four-hour-day creation, but remain reserved in their conclusions of how it all occurred. While all Bibles tell the same basic story, not all Bibles are accurately translated. Most people have never read the entire Bible cover-to-cover even though they claim they have. There is a difference between having read the Bible and having read the *entire* Bible. Sadly, many scientists who reject the Bible grew up with reading a Bible version that is very inappropriate for scientific study, and in some cases it was only a children's Bible they read in their youth. Many professors who dash the faith of young students to pieces know little or nothing about the finer details of the subject, but they know enough to make six-twenty-four-hour-day creationists look like fools during discourse of debate in class.

Do not let your passions dictate your logic. Logic and reasoning should be your passion, thus allowing you to be more likely able to accurately assess "The Science of God", or to be more clear—God's Science. Do not allow yourself to imagine that you are being logical by rejecting the thought of there being a Creator, because rejection without proof is nothing more than ill-directed passion. On the face, without any bogus scientific imaginings by anti-god zealots, the evidence of a Creator is overwhelming. It is unwise to disregard a Creator, to say the least. This does not mean you must accept the idea of a Creator, but rather that you are wise to *not* disregard such.

If we take our minds into the infinite past, we must ask ourselves what was there before–was there anything? And if we imagine to see anything being there then we must ask, how did it get there?

We cannot avoid God and then proceed to concoct some idiotic theory that beings from some other realm or universe made it all, because that still does not explain how *they* would have come to be. In doing so, no matter how hard you try to avoid it, you will always be chasing the proverbial carrot hanging on the stick, thus inventing one new ridiculous theory after another.

Always take yourself back to a vast empty space that is void of all things and start from there in order to begin to understand how everything came to be. If you have a reasonable mind you should be able to deduce that anything physical is *something* and it had to somehow get there during the infinite past. How did it get there? And once it was there, what sequence of events might have occurred to produce all that we see today? Let your mind be open and true.

A Summary

To summarize a few key points, realize that "heaven and earth" in the first sentence of Genesis One are not the same as some of the later references of those terms, and that is your primary key to begin understanding the Genesis One text.

Words like "firmament" are critical to grasp, and they should never be replaced with human-view terms like "expanse", "canopy", "vault", "sky", "dome", or even "horizon". etc.

"Light" is not our Sun, and days could not be counted until the fourth "day", thus allowing the previous four days to be legitimately unlimited in time, presenting a view of Genesis that is perfectly in line with physical reality.

If these easily discernable key points are overlooked, you will inevitably misinterpret the Genesis One text and be lured into incorrect analysis of that text. Think critically and dispose of hocus-pocus magical beliefs of instant things like big bang and six-twenty-four-hour creation. Proper authoritative Bible versions of Genesis One are solid, undeniable, scientific documents. Accept them as such and you will see Truths that you never anticipated!

A Scientific Prayer

If we are going to *believe*, then we should pray for guidance and it should go something like this:

Oh wondrous and glorious self-Created Creator, Your Creation is without bounds. The deeper we peer the more we see. The more we see, the more we realize we do not understand Your ways. You are from everlasting to everlasting. Your Creation cannot be dated by scientific reasoning. We can barely conceive the billions of years we demand that we limit You to in our minds, let alone the truth of the matter that likely exceeds that time.

Your mastery of Your elements will thrill us for ages as we try to disprove Your gracious care.

We speak to You as representatives of humanity. Pardon our arrogance of imagining that we can see the extents of Your visible Creations. Those with sense understand that if we peer yet deeper into the extents of the heavens, we will see more of what we now see. And when we achieve that and decide to peer yet deeper, we again will see more of the same of Your infinite Creation and nature. In our arrogance, we foolishly demand that you do not exist, but then we claim everything simply popped into existence of its own accord—Our arrogance is never ending. Forgive us for our poor use of the language that You have so graciously given us and forgive us for how we have ignorantly perverted the brief logical account of Creation that You have so graciously shared with us in Genesis One.

Deal with us, oh wondrous Creator, all according to our deeds, an Eye for an Eye, as we either destroy or increase others with our lies or with our truths. Pay us according to our works. As we discredit the gift of salvation for our own personal gain, pay us according to our lies. And as we guide those to true understanding, pay us according to our truths.

The splendor of Your heavens give testament to Your undeniable existence. We foolishly seek to scientifically detect Your existence and then arrogantly believe that You do not exist because Your intangible nature cannot be sensed through machines, but rather is only knowable through the mysterious mind of man.

Forgive our foolish ways as we forsake Your gracious Word. We speak, but we do not understand the words that flow forth from our mouths. We continually credit the wrong source for our understanding, and through our foolishness, we forfeit True Understanding.

Purify our hearts and minds and make us anew with new understanding using Your truth as our logic and reasoning, not pretending and inventing as we go along, but rather finding Your Truth to understand and know You better. Help us to think clearly and bring the proper people and their minds into each of our lives to strengthen and clarify our own understanding, thus drawing us ever closer to Truth and to what is actually true.

Chasten us when we veer from Your straight path of truth. Discipline us when we lead others astray. Teach us what is right and what is wrong, for we have perverted our minds to believe folly and have lost our way.

Show us the way, oh wondrous Creator, to know that Creation is Yours. And if that way sits before our very own eyes, then heal us from our blindness that we may see and tell all the world of Your existence through...

The Science of God

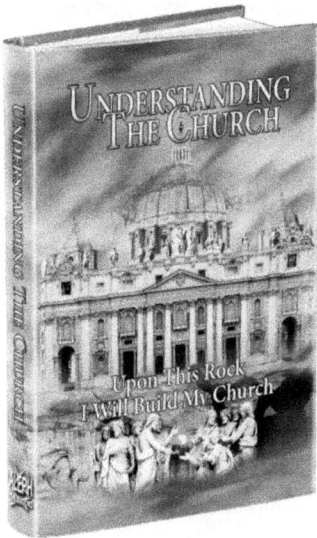

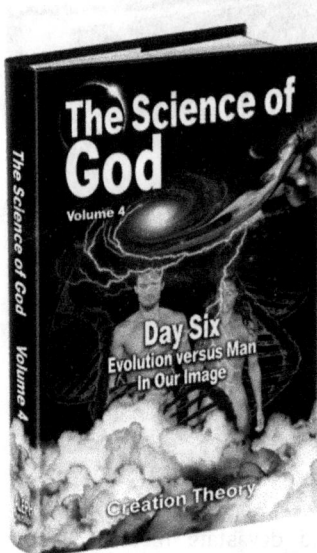

The Science of God
Volume 3
Day Five and Day Six
The Creatures – Revolution or Evolution

A Fishy Flying Account Crawling Out of Nowhere

Have you been trying to share your views in the evolution versus creation debate but are thwarted at your every utterance? Are you reluctant to speak up and share your opinions because you're not sure what is or is not true? Sometimes we might even wonder if we should even bother pondering these things at all since no one can ever truly prove their theory to a point of it being "undeniable fact".

Take heart because there are more possibilities than are offered by most people on either side of the discussion. Bystanders often observe the views from both sides of the debate and will then consider those perspectives and try to balance them using logic, but we often fail to achieve that logical balance.

Balance is achieved by many people, but it is typically compromised in order to arrive at an agreeable viewpoint. Ignoring facts in this way is no way to discover truth.

The Science Of God Volume 3 – The Creatures – Revolution or Evolution will not force you to ignore any true facts, and will guide you on your quest to see the clear path to how creatures came to be. God or Evolution? You decide, because everyone is welcome in the discussion!

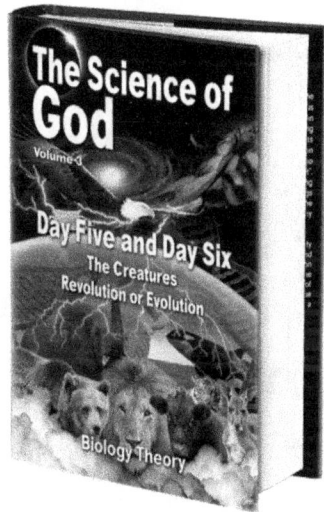

Search: The Science Of God Book Volume 3
SayItBooks.com

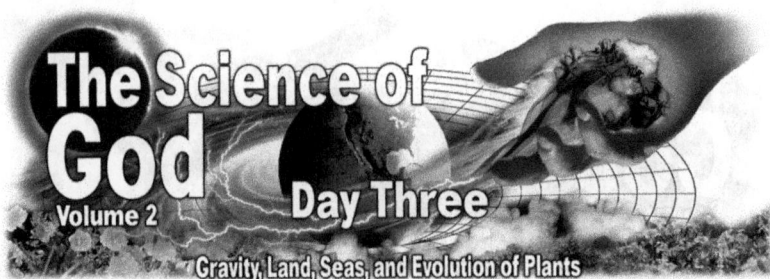

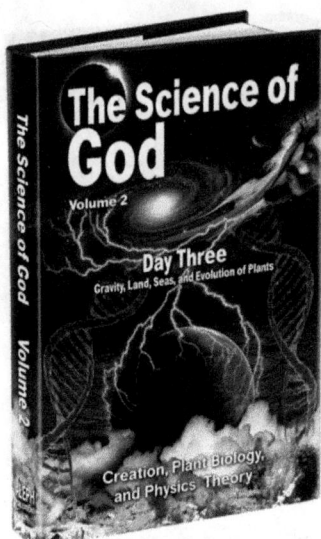

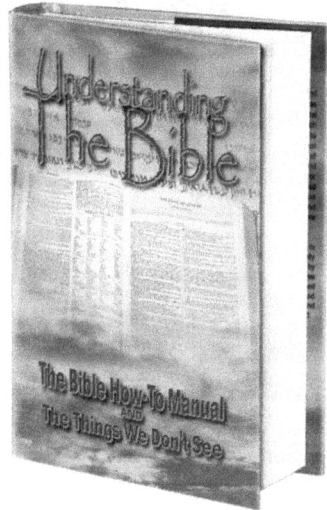

The Prayer How-To Manual
Understanding
Prayer
Why Our Prayers Don't Work

Learn the Real Secret of Prayer

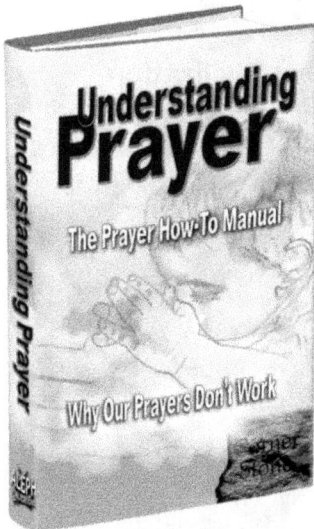

There's a secret that many have tried to understand but failed to accomplish. We pray day after day after day with little or no positive results, causing us to lose faith.

Some people believe that there's a secret method that must be followed to get your prayers answered and receive the things you want in life, but their success is limited, if it comes at all; while others believe that they're not worthy to have their prayers answered. Few people know the True secret, and when they tell us we often misunderstand them.

Understanding Prayer explains, in easy to grasp language, the mysteries behind many causes of prayer failure. True success in your prayers is not measured by how often you pray, how long you pray, or even how badly you want something and how hard you for pray it. True success in your prayer life is measured by *results*!

Understanding Prayer offers you the opportunity to get those results as it reveals the mysteries of a full and robust prayerful connection allowing you solid and repeatable results nearly on command. A little time to read and pray is all it takes to quickly put these sound, true, simple principles to work for you and your family. Gain the understanding of prayer and of how to receive the blessings of financial and mental wealth that can benefit you and keep you free from strife and trouble for years to come!

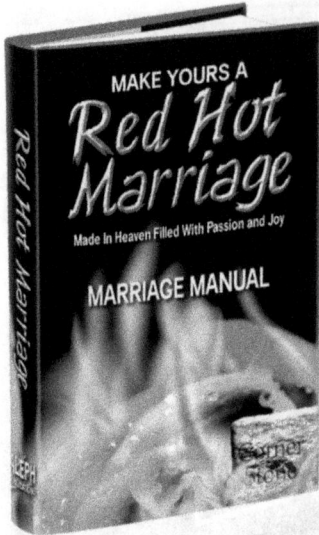

Notes

Notes

Notes

Notes